Power Systems

Electrical power has been the technological foundation of industrial societies for many years. Although the systems designed to provide and apply electrical energy have reached a high degree of maturity, unforeseen problems are constantly encountered, necessitating the design of more efficient and reliable systems based on novel technologies. The book series Power Systems is aimed at providing detailed, accurate and sound technical information about these new developments in electrical power engineering. It includes topics on power generation, storage and transmission as well as electrical machines. The monographs and advanced textbooks in this series address researchers, lecturers, industrial engineers and senior students in electrical engineering.

Power Systems is indexed in Scopus

Dariusz Czarkowski • Qijun Deng • Erdem Asa

Converter Topologies for Resonant Wireless Power Transfer

Springer

Dariusz Czarkowski
Electrical and Computer Engineering
University of South Alabama
Mobile, AL, USA

Qijun Deng
School of Electrical Eng. and Automation
Wuhan University
Wuhan, China

Erdem Asa
Oak Ridge National Laboratory
Oak Ridge, TN, USA

ISSN 1612-1287 ISSN 1860-4676 (electronic)
Power Systems
ISBN 978-3-032-31447-5 ISBN 978-3-032-31448-2 (eBook)
https://doi.org/10.1007/978-3-032-31448-2

This Springer imprint is published by the registered company Springer Nature Switzerland AG
The registered company address is: Gewerbestrasse 11, 6330 Cham, Switzerland

To our families.

Acknowledgement The contribution to this work from Oak Ridge National Laboratory, supported by the U.S. Department of Energy, is gratefully acknowledged.

Competing Interests The authors have no competing interests to declare that are relevant to the content of this manuscript.

Contents

About the Authors

Dariusz Czarkowski earned his master's degree in electrical engineering from AGH University of Krakow, Poland, and Wright State University in Dayton, Ohio, USA, and completed his Ph.D. in electrical engineering at the University of Florida in Gainesville, FL, USA. He was honored with Professor Emeritus status at the New York University Tandon School of Engineering, Brooklyn, NY, USA, an institution he served for nearly three decades. He is currently a part-time instructor in the Department of Electrical and Computer Engineering at the University of South Alabama. He has published extensively on power electronics and power distribution systems. He is the co-author of the seminal textbook *Resonant Power Converters*. His research and teaching continue to focus on the design and advanced control of resonant and PWM power converters using data-driven methods.

Qijun Deng received B.S. and M.Sc. degrees in mechanical engineering and Ph.D. degree in computer application technology from Wuhan University, Wuhan, China, in 1999, 2002, and 2005, respectively. In 2005, he joined Wuhan University, where he is currently a Professor in the School of Electrical Engineering and Automation. He has authored or co-authored over 100 journal and conference publications. From 2013 to 2014, he was a Visiting Scholar with the New York University Tandon School of Engineering, Brooklyn, NY, USA. His research interests include wireless power transfer, distribution automation, and electrical power informatics.

Erdem Asa received his Ph.D. degree in electrical engineering from the New York University Tandon School of Engineering, Brooklyn, NY, USA, in 2016. He is currently an R&D Staff Member with the Vehicle Power Electronics Research Group at Oak Ridge National Laboratory. He has authored or co-authored over 40 journal and conference publications. Before joining Oak Ridge National Laboratory in 2019, he was a Design and Control Power Electronics Engineer with GE Aviation. His current research interests include isolated and non- isolated dc–dc converters, power converters for the aviation industry, plug-in hybrid and electric vehicles, and wireless power transfer systems.

Abbreviations

AGV	Automated guided vehicle
AUV	Autonomous underwater vehicle
BMS	Battery management system
CC/CV	Constant current/constant voltage
CCM	Continuous conduction mode
CPT	Capacitive power transfer
CT	Current transformer
DWPT	Dynamic wireless power transfer
EMC	Electromagnetic compatibility
EMF	Electromagnetic field
EMI	Electromagnetic interference
ESR	Equivalent series resistance
EV	Electric vehicle
FB	Full bridge
FHA	Fundamental harmonic approximation
FOD	Foreign-object detection
GaN	Gallium nitride
G2V/V2G	Grid-to-vehicle/vehicle-to-grid
HB	Half bridge
HV	High voltage
ICNIRP	International Commission on Non-Ionizing Radiation Protection
ICT	InterCell transformer
ISM	Industrial, scientific, and medical (frequency band)
KCL / KVL	Kirchhoff's current/voltage law
LCL, LCC	Inductor–capacitor–inductor/inductor–capacitor–capacitor compensation
LOD	Living-object detection
MOSFET	Metal-oxide-semiconductor field-effect transistor
MPP	Magnetic Power Profile (Qi2)
NPC	Neutral-point clamped

OMTD	Online monitoring terminal device
PFC	Power factor correction
P-P, P-S, S-P, S-S	Parallel–parallel, parallel–series, series–parallel, series–series compensation
PWM	Pulse-width modulation
SAE	Society of Automotive Engineers
SiC	Silicon carbide
THD	Total harmonic distortion
UAV	Unmanned aerial vehicle
WBG	Wide bandgap
WPC	Wireless Power Consortium
WPT	Wireless power transfer
ZCS	Zero-current switching
ZDS	Zero-derivative switching
ZVS	Zero-voltage switching

Chapter 1
Introduction

Wireless power transfer (WPT) has evolved from early experimental demonstrations into a broad family of technologies spanning consumer electronics, medical implants, industrial systems, and electric-vehicle charging. This introductory chapter places resonant WPT in a historical and technical context by reviewing major milestones, application domains, standardization efforts, and present-day design constraints. The discussion distinguishes near-field and far-field approaches, summarizes representative frequency ranges and usage cases, while highlighting the roles of interoperability, electromagnetic compatibility, exposure limits, and safety functions in practical systems. A generalized system architecture is then introduced, together with circuit, equivalent-network, and multi-coil modeling viewpoints that recur throughout the monograph. These perspectives support the central theme of this monograph: converter topologies, compensation networks, rectifiers, and control strategies must be selected and designed as an integrated resonant power-electronic system rather than as isolated subsystems.

1.1 Past and Present of WPT

Wireless power transfer (WPT) encompasses technologies that deliver electrical energy across a gap without physical contact. These include near-field approaches (predominantly magnetic or electric coupling) and far-field methods (radiative power beaming). This monograph focuses on *converter topologies for resonant WPT* as used in consumer, industrial, medical, and transportation systems. The intent of this section is to (i) provide historical context for how WPT evolved into a set of modern, standardized product categories, and (ii) motivate why resonant converters and compensation networks are central to present-day WPT engineering.

D. Czarkowski et al., *Converter Topologies for Resonant Wireless Power Transfer*, Power Systems, https://doi.org/10.1007/978-3-032-31448-2_1

1.1.1 Classification: Coupling Mechanism, Range, and Power

Engineering trade-offs in WPT are best characterized by three attributes: coupling mechanism, transfer distance, and delivered power. Near-field methods rely on non-radiative coupling and typically maximize efficiency at short distances:

- **Tightly-coupled Resonant WPT:** Relies on high coupling coefficients ($k > 0.1$) and operates at frequencies such as 110–205 kHz (Qi) or 360 kHz (Qi2 MPP) to maximize power density and efficiency in consumer electronics.
- **Loosely-coupled Resonant WPT:** Uses high-quality-factor (Q) resonators to maintain efficiency across large air gaps and significant misalignments ($k < 0.1$), typified by the 85 kHz band defined by SAE J2954 for electric vehicles and industrial AGVs (20–100 kHz).
- **High-Frequency Magnetic Resonance:** Operates in the MHz range (e.g., 6.78 or 13.56 MHz ISM bands) to provide "spatial freedom," enabling multi-device charging and power transfer to miniaturized medical implants.
- **Capacitive WPT (CPT):** Employs electric-field coupling between conductive plates; attractive in metal-rich environments or applications involving rotating shafts where magnetic fields may induce unwanted eddy currents.

While standards like SAE J2954 and Qi provide interoperability within fixed bands, custom industrial and medical resonant systems often operate at proprietary frequencies (ranging from 20 kHz to several MHz) tailored to specific constraints of efficiency, volume, and biological exposure limits. Far-field WPT uses radiating antennas and rectennas, offering high spatial flexibility at the expense of regulatory and safety constraints, and typically requiring careful beam control [1]. Table 1.1 summarizes common WPT families, typical frequency bands, and representative application domains. The automotive EV domain is increasingly anchored around

Table 1.1 Typical frequency bands and applications for representative WPT regimes

Coupling regime	Frequency range	Standards/typical applications
Tightly-coupled	80–360 kHz	**Qi:** 110–205 kHz; **Qi2 (MPP):** 360 kHz. High-efficiency charging for portable electronics.
Loosely-coupled	20–150 kHz	**SAE J2954 (EV):** 85 kHz; **Industrial:** 20–100 kHz. AGVs and high-power transport systems.
High-Freq. Res.	6.78 MHz, 13.56 MHz	**AirFuel:** Spatial freedom and multi-device charging; high-frequency medical implants.
Capacitive (CPT)	100 kHz to 13.56 MHz	Electric-field coupling; ideal for metal-rich environments or rotating shafts.
Microwave	2.4 GHz, 5.8 GHz	Radiative far-field; long-range beaming, RFID, and IoT sensor nodes.

an 85 kHz nominal band motivated by interoperability and electromagnetic compatibility considerations [2–4]. Although microwave WPT is listed in Table 1.1 for completeness, this monograph focuses on converter topologies for near-field resonant WPT, and therefore does not treat far-field microwave power-beaming circuits.

Beyond this taxonomy, modern WPT systems are best understood as *power-electronic systems* rather than isolated couplers. Converter choice determines the achievable waveform family (sinusoidal vs. quasi-square excitation), the feasible operating frequency, and the controllable degrees of freedom (frequency, phase, duty ratio, and multi-phase modulation), all of which shape efficiency, thermal behavior, electromagnetic compatibility, and safety testing.

1.1.2 Historical Roots: Tesla's Experiments and the "Wireless Energy" Vision

The history of "wireless power" is traced to Nikola Tesla's late-nineteenth-century experiments and patents describing systems intended for long-distance energy transfer without wires [5]. The chronology of WPT (Fig. 1.1) highlights a transition from the bold, high-voltage visions of the early twentieth century (Fig. 1.2) to the highly efficient, standardized resonant devices used in modern systems (Fig. 1.3). Tesla's Colorado Springs work (1899–1900) and the Wardenclyffe follow-up (1901–1904) established the idea that high-voltage, high-frequency excitation could enable unconventional transmission paths and large electric fields [6]. While modern resonant WPT does *not* follow Tesla's earth-resonance vision, his work remains a useful historical reference point: it illustrates that wireless energy transfer must be treated as a system, encompassing source, coupling medium, receiver, and safety constraints.

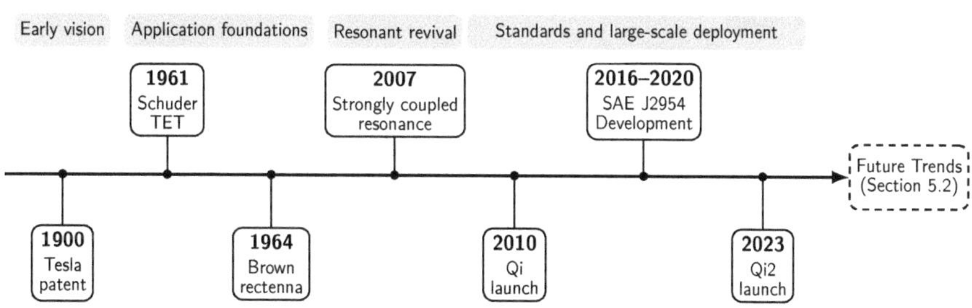

Fig. 1.1 A compact timeline of landmark developments that shaped modern WPT (selected milestones)

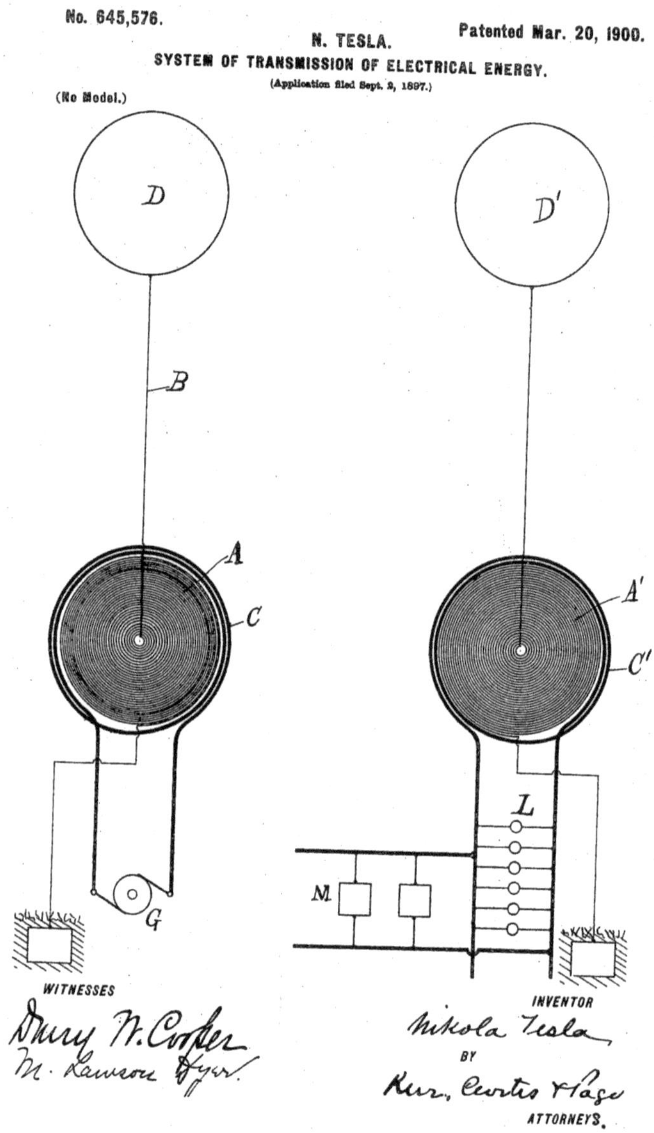

Fig. 1.2 Facsimile of the foundational 1900 patent (US 645,576) by Nikola Tesla. The schematic illustrates the early concept of resonant wireless energy transmission using elevated terminals and grounded transformers (Source: U.S. Patent and Trademark Office)

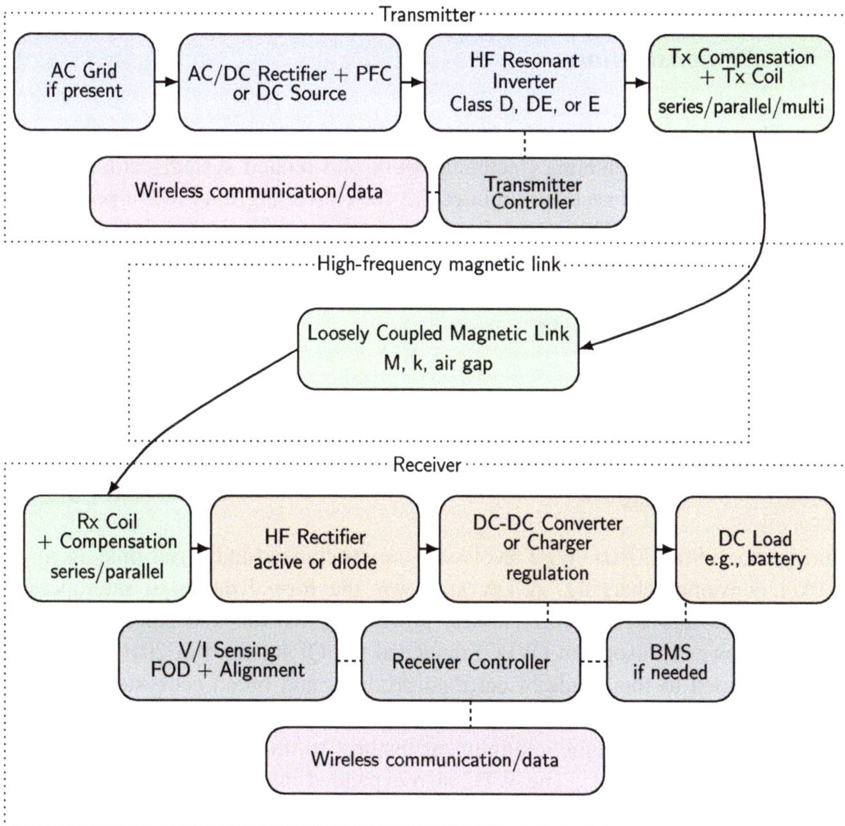

Fig. 1.3 Generalized architecture of a resonant WPT system

1.1.3 Medical Implants: Transcutaneous Energy Transfer as an Early Flagship Application

One of the earliest application-driven areas of WPT development was implantable medical devices. In 1961, energy transfer into a closed chest using stationary coupling coils and a portable high-power oscillator was reported [7]. This work laid the foundation for the transcutaneous energy transformer concept, with continued emphasis on reliable alignment, tissue heating limits, and fail-safe operation [8]. Modern implantable WPT remains a distinct design space: power levels can be modest compared to EV charging, but the constraints on coupling geometry, efficiency, and biological exposure are far more stringent [9].

1.1.4 Far-Field Power Beaming: The Rectenna and Microwave Demonstrations

The mid-twentieth century also saw rapid progress in microwave components and antennas. Rectifying antenna (rectenna) work and related system demonstrations established a practical receiver architecture for converting microwave power back to DC efficiently [1]. This technology became a cornerstone for power-beaming concepts (e.g., solar power satellites) and continues to influence modern research on long-range WPT where directional transmission is both technically feasible and subject to regulatory oversight.

1.1.5 Industrial and Consumer WPT: From Inductive Links to Global Standards

During the 1990s–2010s, WPT evolved from specialized industrial links to mass-market consumer charging. A key step was the formalization of interoperable specifications that define coil classes, power negotiation, and safety tests. The Wireless Power Consortium (WPC) launched the Qi standard in 2010, and it has since expanded to thousands of certified products and broad ecosystem adoption [10]. The Qi specification has also evolved: Qi v2.0 (branded as Qi2) was launched in 2023, introducing magnetic alignment features to improve user experience and energy efficiency [10, 11]. The WPC also expanded into higher-power appliances with the Ki standard for cordless kitchen appliances (up to 2200 W), illustrating how standardization can open entirely new WPT product categories [12]. Safety and robustness features became primary design requirements as WPT entered everyday use. Foreign-object detection (FOD) and living-object protection are now core elements of many systems; a modern technical overview of detection approaches is provided in [13].

1.1.6 The Resonance Resurgence: Strongly Coupled Magnetic Resonances and Spatial Freedom

A widely cited catalyst for renewed academic and industrial interest in mid-range magnetic WPT was a 2007 *Science* paper demonstrating transfer via strongly coupled magnetic resonances [14]. This line of work accelerated research into multi-coil systems, impedance matching under misalignment, and the co-design of resonators with efficient power electronics. Industry consortia such as AirFuel promoted resonance-based charging aimed at multi-device and "spatial freedom" experiences, including operation at higher ISM-band frequencies (e.g., 6.78 MHz) [15].

1.1.7 High-Power WPT for Electric Vehicles: Interoperability, EMF, and the 85 kHz Band

EV charging occupies the upper end of near-field WPT in terms of both power level and air gap. High delivered power, large air gaps, and human proximity drive stringent electromagnetic field (EMF) management and safety validation. The SAE J2954 standardization effort provides an interoperability and alignment framework for light-duty vehicles [4]. Technical reviews of high-power WPT commonly discuss the rationale for a nominal 85 kHz operating band [2], and surveys of static and dynamic EV charging provide a broader perspective on infrastructure and grid integration [3, 16].

1.1.8 Present-Day Design Constraints: Regulation, Exposure Limits, and Converter Implications

Modern WPT must be designed within exposure and EMC limits. ICNIRP provides widely referenced guidelines for limiting human exposure to time-varying electric and magnetic fields, including the low-frequency range relevant to many inductive systems [17]. IEEE Std C95.1 also defines exposure criteria across a very wide frequency range [18]. These limitations influence coil geometry, shielding, allowable leakage fields, and ultimately power electronics choices (e.g., switching frequency, modulation strategy, and control bandwidth). From the converter-topology standpoint, the present landscape can be summarized as:

- **Efficiency under variable coupling and load:** modern systems require high efficiency over misalignment and varying receiver states, which motivates soft-switching resonant inverters and adaptive matching [19];
- **High-frequency operation with wide bandgap devices:** GaN/SiC devices enable higher switching frequencies and reduced passive size, but intensify EMC and gate-drive design requirements [19];
- **Safety-driven sensing and control:** FOD, thermal limits, and exposure compliance necessitate sensing, estimation, and closed-loop power control integrated with the inverter and rectifier [13].

Today, WPT is shaped less by any single coupling mechanism than by three converging factors: mature standards (Qi/Qi2, SAE J2954), safety and EMC constraints (ICNIRP/IEEE), and power-electronics advances that make high-efficiency resonant conversion practical across wide operating ranges. These constraints define the focus on converter topologies in this monograph, where resonant inverters, rectifiers, and compensation networks are developed in a cohesive design space.

1.2 System Structure

Figure 1.3 shows the overall architecture of a resonant WPT system. Although Fig. 1.3 shows a more common magnetic link, the same architecture applies to capacitive WPT, where the coils and mutual inductance are replaced by coupled plates and a coupling capacitance. A complete implementation typically contains a front-end rectifier and resonant inverter on the primary side, a magnetic coupler formed by the transmitting and receiving coils (or a capacitive coupler), and a receiver-side rectifier followed by one or more DC–DC conversion stages. Practical systems also incorporate voltage and current sensing, charge control, alignment assistance, foreign-object detection, and communication functions [20, 21]. Depending on the application, the grid-side front end may also include power-factor correction to reduce apparent power draw from the utility source.

In circuit terms, the coupler behaves as a loosely coupled transformer with a comparatively large separation between the windings. This geometry increases leakage inductance and winding losses, while reducing magnetizing inductance and mutual inductance relative to tightly coupled transformers. For that reason, additional compensation capacitors are normally required on both the transmitter and receiver sides to form resonant tanks and to shape the effective source and load seen by the power converters. In modern WPT systems, double-sided compensation is generally preferred over single-sided compensation because it offers more degrees of freedom for satisfying efficiency, controllability, and soft-switching requirements [22].

Compensation-topology selection is guided by four system-level requirements [22]:

(1) **Minimized VA rating and maximized power-transfer capability.** Compensation should cancel the reactive part of the coil impedances so that the source VA rating is reduced on the primary side and the transfer capability is improved on the secondary side.
(2) **Constant-voltage or constant-current behavior.** Proper topology choice can provide desirable output characteristics with reduced control burden, which is especially valuable under coupling and load variation.
(3) **High efficiency.** Besides improving power transfer, compensation strongly affects the input phase angle and therefore the ability of the inverter to achieve soft switching.
(4) **Bifurcation resistance and robustness.** The chosen network should avoid ambiguous zero-phase-angle operating points and should remain well behaved under parameter variation, misalignment, and, when required, bidirectional operation.

Figure 1.4 shows the model family used throughout this monograph. In the basic circuit model, the transmitting coil L_1 and receiving coil L_2 are linked by mutual inductance M_{12}, while R_1 and R_2 represent the equivalent ac resistances of the two coils. The same system can be recast as an equivalent T-model or as a two-port

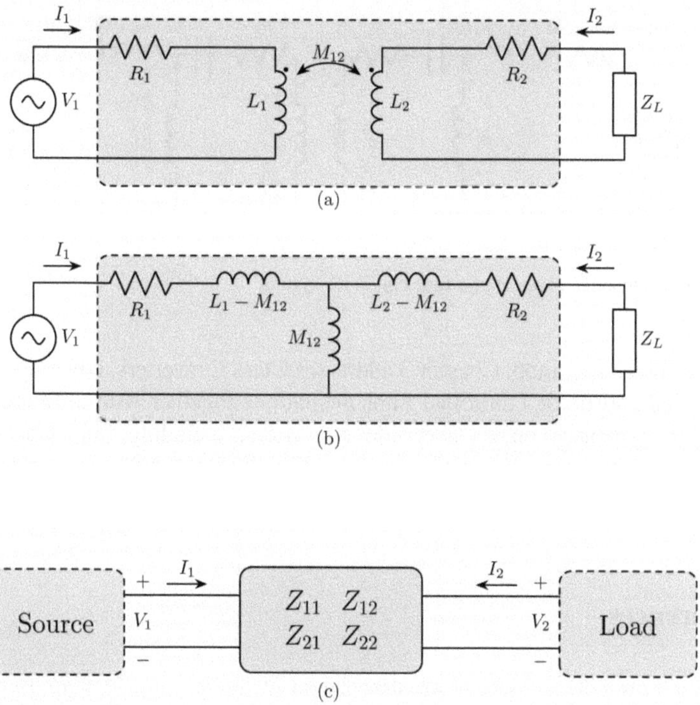

Fig. 1.4 A typical WPT system viewed at the coupler and network level: (**a**) circuit model, (**b**) equivalent T-model, and (**c**) two-port-network model

network, depending on whether circuit intuition or input–output characterization is more convenient for the design task [23]. Figure 1.4b is the T-equivalent of the coupled-coil pair expressed directly in terms of L_1, L_2, and M_{12}; no separate ideal transformer or explicit turns ratio is introduced.

Not all practical systems contain every functional block shown in Fig. 1.3. For example, some battery-powered systems do not require a separate grid-side power-factor-correction stage, whereas other implementations may include additional sensing, shielding, or communications hardware. Likewise, multi-coil WPT structures can be introduced to improve transmission efficiency, misalignment tolerance, or spatial freedom.

Figure 1.5 shows one such extension: a four-coil system in which auxiliary sending and receiving coils are added between the source/load and the main power-transfer coils. In that case, additional parasitic capacitances and resistances must be taken into account, and the designer gains additional tuning flexibility at the expense of increased component count and modeling complexity.

The rest of this monograph is organized as follows. Chapter 2 develops Class D compensated inverters, including series–series compensation, other compensation schemes, multi-element resonant inverters, and multiphase inverters for

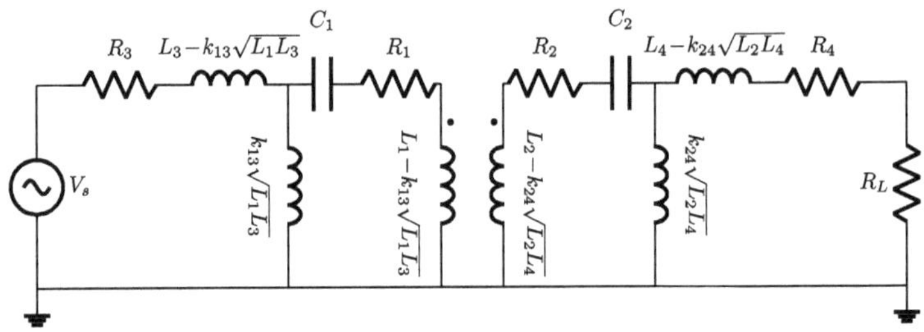

Fig. 1.5 Equivalent T-model of a four-coil WPT system

higher-power operation. Chapter 3 addresses Class E inverters, covering frequency-controlled and phase-controlled implementations together with a design example. Chapter 4 focuses on receiver-side converters, including impedance-matching circuits, rectifiers, and DC–DC converters. Finally, Chap. 5 discusses representative applications and selected future trends in WPT systems.

References

1. W.C. Brown, The history of the development of the rectenna, in *Solar Power Satellite Microwave Power Transmission and Reception*, NASA Conference Publication 2141 (NASA Lyndon B. Johnson Space Center, Houston, 1980), pp. 271–280. Accessed 26 Mar 2026 [Online]. Available: https://adsabs.harvard.edu/full/1980NASCP2141..271B
2. A. Foote, O.C. Onar, A review of high-power wireless power transfer, in *2017 IEEE Transportation Electrification Conference and Expo (ITEC)* (2017), pp. 234–240. https://doi.org/10.1109/itec.2017.7993277
3. C. Panchal, S. Stegen, J. Lu, Review of static and dynamic wireless electric vehicle charging system. Eng. Sci. Technol. Int. J. **21**(5), 922–937 (2018). https://doi.org/10.1016/j.jestch.2018.06.015
4. SAE International, *Wireless Power Transfer for Light-Duty Plug-in Electric Vehicles and Alignment Methodology* (SAE International, Warrendale, 2024). https://doi.org/10.4271/j2954_202408
5. N. Tesla, System of transmission of electrical energy, US645576A, Filed 1897; granted March 20, 1900. Accessed 26 Mar 2026 [Online]. Available: https://patents.google.com/patent/US645576A/en
6. PBS, Tesla – life and legacy: Colorado springs. Accessed 16 Apr 2026 [Online]. Available: https://www.pbs.org/tesla/ll/ll_colspr.html
7. J.C. Schuder, H.E.J. Stephenson, J.F. Townsend, Energy transfer into a closed chest by means of stationary coupling coils and a portable high-power oscillator. Trans. Am. Soc. Artif. Internal Organs **7**, 327–331 (1961). Accessed 26 Mar 2026 [Online]. Available: https://pubmed.ncbi.nlm.nih.gov/13748744/
8. J.C. Schuder, Powering an artificial heart: Birth of the inductively coupled-radio frequency system in 1960. Artif. Organs **26**(11), 909–915 (2002). https://doi.org/10.1046/j.1525-1594.2002.07130.x
9. S.R. Khan, S.K. Pavuluri, G. Cummins, M.P.Y. Desmulliez, Wireless power transfer techniques for implantable medical devices: a review. Sensors **20**(12), 3487 (2020). https://doi.org/10.3390/s20123487

10. Wireless Power Consortium, History of the Qi specifications. Accessed 26 Mar 2026 [Online]. Available: https://www.wirelesspowerconsortium.com/knowledge-base/qi-specification/history-of-the-qi-specifications/
11. Wireless Power Consortium, WPC to build on successful Qi2 rollout, announces pending Ki standard launch. Accessed 26 Mar 2026 [Online]. Available: https://www.businesswire.com/news/home/20240104519170/en/WPC-To-Build-On-Successful-Qi2-Rollout-Announces-Pending-Ki-Standard-Launch
12. Wireless Power Consortium, Ki cordless kitchen. Accessed 20 Mar 2026 [Online]. Available: https://www.wirelesspowerconsortium.com/standards/ki-cordless-kitchen/
13. J. Lu, G. Zhu, C.C. Mi, Foreign object detection in wireless power transfer systems. IEEE Trans. Ind. Appl. **58**(1), 1340–1354 (2022). https://doi.org/10.1109/tia.2021.3057603
14. A. Kurs, A. Karalis, R. Moffatt, J.D. Joannopoulos, P. Fisher, M. Soljačić, Wireless power transfer via strongly coupled magnetic resonances. Science **317**(5834), 83–86 (2007). https://doi.org/10.1126/science.1143254
15. AirFuel Alliance, Frequency choice. Accessed 26 Mar 2026 [Online]. Available: https://airfuel.org/frequency-choice/
16. Z. Xue, W. Liu, C. Liu, K.T. Chau, Critical review of wire- less charging technologies for electric vehicles. World Electric Veh. J. **16**(2), 65 (2025). https://doi.org/10.3390/wevj16020065
17. International Commission on Non-Ionizing Radiation Protection, Guidelines for Limiting Exposure to Time-Varying Electric and Magnetic Fields (1 Hz to 100 kHz). Health Phys. **99**(6), 818–836 (2010). https://doi.org/10.1097/hp.0b013e3181f06c86
18. IEEE, *IEEE Standard for Safety Levels with Respect to Human Exposure to Electric, Magnetic, and Electromagnetic Fields, 0 Hz to 300 GHz*. Institute of Electrical and Electronics Engineers (2019). https://doi.org/10.1109/ieeestd.2019.8859679
19. J.V. Mulders, D. Delabie, C. Lecluyse, C. Buyle, G. Callebaut, L.V. der Perre, L.D. Strycker, Wireless power transfer: systems, circuits, standards, and use cases. Sensors **22**(15), 5573 (2022). https://doi.org/10.3390/s22155573
20. Y. Gao, C. Duan, A.A. Oliveira, A. Ginart, K.B. Farley, Z.T.H. Tse, 3-D coil positioning based on magnetic sensing for wireless EV charging. IEEE Trans. Transp. Electrif. **3**(3), 578–588 (2017). https://doi.org/10.1109/tte.2017.2696787
21. A. Triviño-Cabrera, J.A. Aguado (eds.), *Emerging Capabilities and Applications of Wireless Power Transfer* (IGI Global, Hershey, 2019). ISBN: 9781522558705. https://doi.org/10.4018/978-1-5225-5870-5
22. W. Zhang, C.C. Mi, Compensation topologies of high-power wireless power transfer systems. IEEE Trans. Veh. Technol. **65**(6), 4768–4778 (2016). https://doi.org/10.1109/tvt.2015.2454292
23. Z. Zhang, H. Pang, A. Georgiadis, C. Cecati, Wireless power transfer—an overview. IEEE Trans. Ind. Electron. **66**(2), 1044–1058 (2019). https://doi.org/10.1109/tie.2018.2835378

Chapter 2
Class D Compensated Inverters

This chapter discusses several common compensated topologies for wireless power transfer, including series–series, parallel–series, series–parallel, parallel–parallel, LCC–series, and multi-phase parallel inverters for high-power applications. Among them, the topologies with parallel or LCC compensation at the primary side show good source-protection characteristics; the input source is not protected in topologies with series compensation at the primary side. Without source protection, the inverter could draw an overcurrent if the secondary-side load is disconnected or the mutual inductance decreases to zero. On the other hand, the series topology is simple in structure and high efficiency is relatively easy to obtain. For the LCC topology, in particular, the current through the sending coil remains nearly constant even when the load or mutual inductance varies. The circuit design methods and power-output characteristics of these topologies are analyzed. To achieve maximum output power, the input impedance of the compensated network is always designed to achieve zero reactive component, meaning that the output power reaches its highest level when the operating frequency equals the self-resonant frequency of the network. Design examples are provided for each topology to guide practical implementation.

2.1 Series-series Compensation

In wireless power transfer systems, Class D voltage-source inverters as well as current-source inverters can be employed. A common topology uses series compensation on both sides. For the sending side, the impedance of the coil is compensated by a series capacitance bank, and the same is done for the receiving side. Both half-bridge and full-bridge Class D inverters are discussed in the following.

D. Czarkowski et al., *Converter Topologies for Resonant Wireless Power Transfer*, Power Systems, https://doi.org/10.1007/978-3-032-31448-2_2

2.1.1 Topology

A typical Class D series–series compensated wireless power transfer topology supplied by a voltage source is shown in Fig. 2.1 [1, 2], where the inverter output is replaced by a square wave in part (c), and the effect of the coupling inductors is represented by two current-dependent voltage sources in parts (d) and (e).

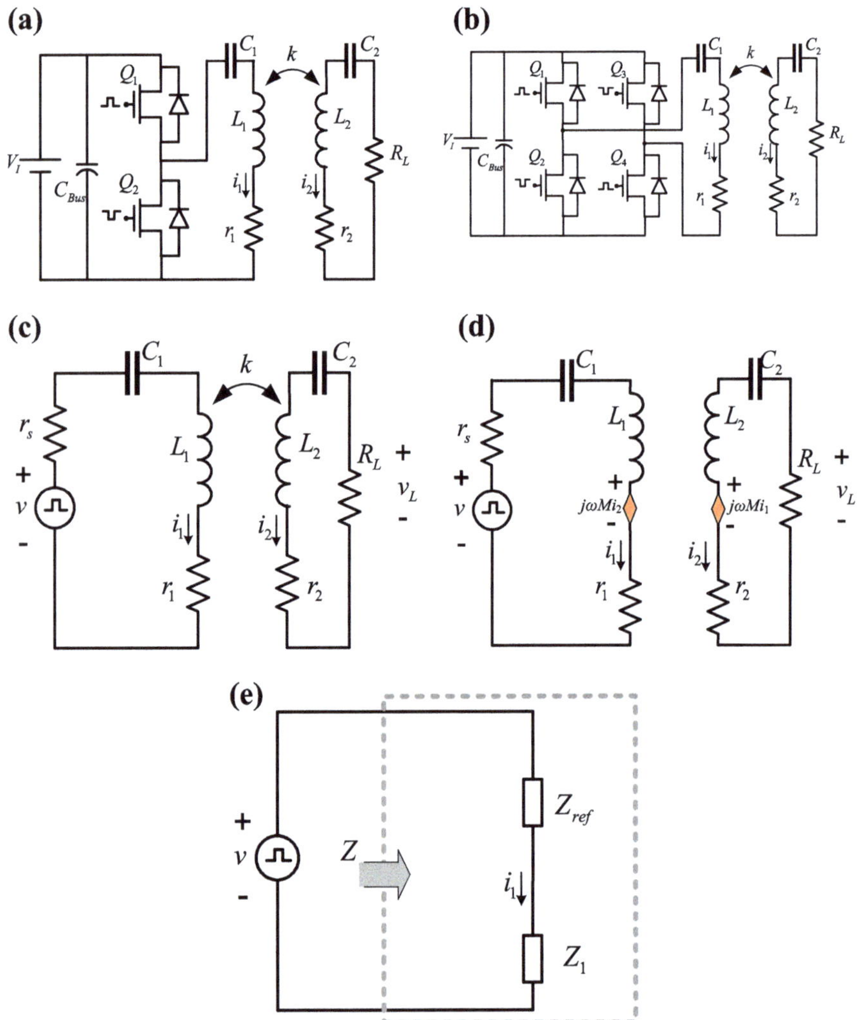

Fig. 2.1 Electrical model of a wireless power transfer system employing a series–series compensation topology: (**a**) half-bridge, (**b**) full bridge, and (**c**)–(**e**) equivalent circuits

In this model, V_I is a DC voltage source, Q_1 through Q_4 are switches (commonly MOSFETs), L_1 and L_2 are two inductors compensated by C_1 and C_2, respectively, r_1 and r_2 are the series resistances of each loop including the equivalent series resistances (ESRs) of the coil and capacitor, r_s is the on-state ESR of the switch, and R_L is the load resistance. The coupling coefficient between the two coils is denoted by k and is expressed as

$$M = k\sqrt{L_1 L_2} \tag{2.1}$$

where M is the mutual inductance between L_1 and L_2. If the duty cycle of the inverter is 0.5, the input voltage of the sending loop from the half-bridge inverter is

$$v = \begin{cases} V_I, & 0 < \omega t \leq \pi, \\ 0, & \pi < \omega t \leq 2\pi. \end{cases} \tag{2.2}$$

The voltage is expanded into a Fourier series as

$$\begin{aligned} v &= \frac{V_I}{2} + \frac{2V_I}{\pi}\sum_{n=1}^{\infty}\frac{1-(-1)^n}{2n}\sin(n\omega t) \\ &= \frac{V_I}{2} + \frac{2V_I}{\pi}\left(\sin\omega t + \tfrac{1}{3}\sin 3\omega t + \tfrac{1}{5}\sin 5\omega t + \cdots\right) \end{aligned} \tag{2.3}$$

2.1.2 Characteristics

The analysis of the WPT system based on Fig. 2.1d is carried out in the frequency domain. One obtains the impedances at the sending side and the receiving side as

$$Z_1 = j\omega L_1 + \frac{1}{j\omega C_1} + r_1 + r_s \tag{2.4}$$

$$Z_2 = j\omega L_2 + \frac{1}{j\omega C_2} + r_2 + R_L \tag{2.5}$$

Applying Kirchhoff's voltage law in the frequency domain,

$$\begin{aligned} v &= Z_1 i_1 + j\omega M i_2, \\ 0 &= j\omega M i_1 + Z_2 i_2. \end{aligned} \tag{2.6}$$

One obtains the currents through the sending loop and the receiving loop as

$$i_1 = \frac{v}{Z_1 + \frac{(\omega M)^2}{Z_2}},$$
$$i_2 = -\frac{j\omega M}{Z_2} i_1 = -\frac{j\omega M}{Z_1 Z_2 + (\omega M)^2} v. \tag{2.7}$$

Defining

$$Z_{\text{ref}} = \frac{(\omega M)^2}{Z_2} \tag{2.8}$$

the current through the sending loop in Eq. (2.7) is transferred into

$$i_1 = \frac{v}{Z_1 + Z_{\text{ref}}} \tag{2.9}$$

which means that the influence of the receiving side can be represented by a reflected impedance whose value is given by Eq. (2.8). As a result, Fig. 2.1d can be transformed into Fig. 2.1e, and the total input impedance at the sending side is

$$Z = Z_1 + Z_{\text{ref}} \triangleq |Z|e^{j\psi} \triangleq |Z|\cos\psi + j|Z|\sin\psi = R + jX \tag{2.10}$$

$$\psi = \arctan\left(\frac{X}{R}\right) \tag{2.11}$$

Equation (2.3) represents the output voltage of the half-bridge inverter. The DC component results in no current because of the series capacitor. Additionally, because of the high impedance of the higher-order harmonic components, only the fundamental component of v,

$$v_{i1} = V_m \sin\omega t \tag{2.12}$$

generates the AC current, while the other components can be ignored, where the amplitude of the fundamental component is given by

$$V_m = \frac{2V_I}{\pi} \tag{2.13}$$

Thus, the rms value of the voltage is

$$V_{\text{rms}} = \frac{V_m}{\sqrt{2}} = \frac{\sqrt{2}V_I}{\pi} \approx 0.45V_I \tag{2.14}$$

Based on Eqs. (2.9) and (2.14), the rms value of the fundamental component current through the sending loop is

$$I_{1,\mathrm{rms}} = \frac{V_{\mathrm{rms}}}{|Z_1 + Z_{\mathrm{ref}}|} \approx \frac{0.45 V_I}{|Z_1 + Z_{\mathrm{ref}}|} \tag{2.15}$$

Hence, the rms value of the fundamental component current through the receiving loop is given by

$$I_{2,\mathrm{rms}} = \frac{\omega M I_{1,\mathrm{rms}}}{|Z_2|} \approx \frac{0.45\omega M V_I}{|Z_2|\,|Z_1 + Z_{\mathrm{ref}}|} \tag{2.16}$$

The transferred power and efficiency are calculated as follows. It is assumed that the circuit is loaded with a pure resistance R_L. The rms value of the voltage across the load and the received power are

$$V_{L,\mathrm{rms}} = I_{2,\mathrm{rms}} R_L = \frac{0.45\omega M V_I R_L}{|Z_2|\,|Z_1 + Z_{\mathrm{ref}}|} \tag{2.17}$$

$$P_2 = \frac{V_{L,\mathrm{rms}}^2}{R_L} = I_{2,\mathrm{rms}}^2 R_L = \left(\frac{0.45\omega M V_I}{|Z_2|\,|Z_1 + Z_{\mathrm{ref}}|}\right)^2 R_L \tag{2.18}$$

Substitution of Eqs. (2.4), (2.5), and (2.8) into Eq. (2.18) yields the transferred power as a function of the operating frequency:

$$P_2 = \frac{(0.45\omega M V_I)^2 R_L}{\left|j\omega L_2 + \dfrac{1}{j\omega C_2} + r_2 + R_L\right|^2} \cdot \frac{1}{\left|j\omega L_1 + \dfrac{1}{j\omega C_1} + r_1 + r_s + \dfrac{(\omega M)^2}{j\omega L_2 + \dfrac{1}{j\omega C_2} + r_2 + R_L}\right|^2} \tag{2.19}$$

For simplicity, one can assume that the resonant components at both sides, including the inductors and capacitors, have the same parameters, namely

$$\begin{aligned} L_1 &= L_2 = L, \\ C_1 &= C_2 = C, \\ r_1 &= r_2 = r. \end{aligned} \tag{2.20}$$

and define the self-resonant angular frequency and unloaded quality factor as

$$\omega_0 = \frac{1}{\sqrt{LC}}, \qquad Q_0 = \frac{\omega_0 L}{r} \tag{2.21}$$

Substitution of Eqs. (2.1), (2.20), and (2.21) into Eq. (2.19) leads to

$$P_2 = \frac{(0.45\,\omega k L V_I)^2 R_L}{\left| r\left(1+\frac{R_L}{r}+jQ_0\left(\frac{\omega}{\omega_0}-\frac{\omega_0}{\omega}\right)\right)\right|^2} \cdot \frac{1}{\left| r\left(1+jQ_0\left(\frac{\omega}{\omega_0}-\frac{\omega_0}{\omega}\right)\right)+r_s+\frac{(\omega k L)^2}{r\left(1+\frac{R_L}{r}+jQ_0\left(\frac{\omega}{\omega_0}-\frac{\omega_0}{\omega}\right)\right)}\right|^2} \tag{2.22}$$

Figure 2.2 depicts the received power P_2 and the impedance angle ψ as functions of the normalized frequency ω/ω_0 under the parameters listed in Table 2.1. The

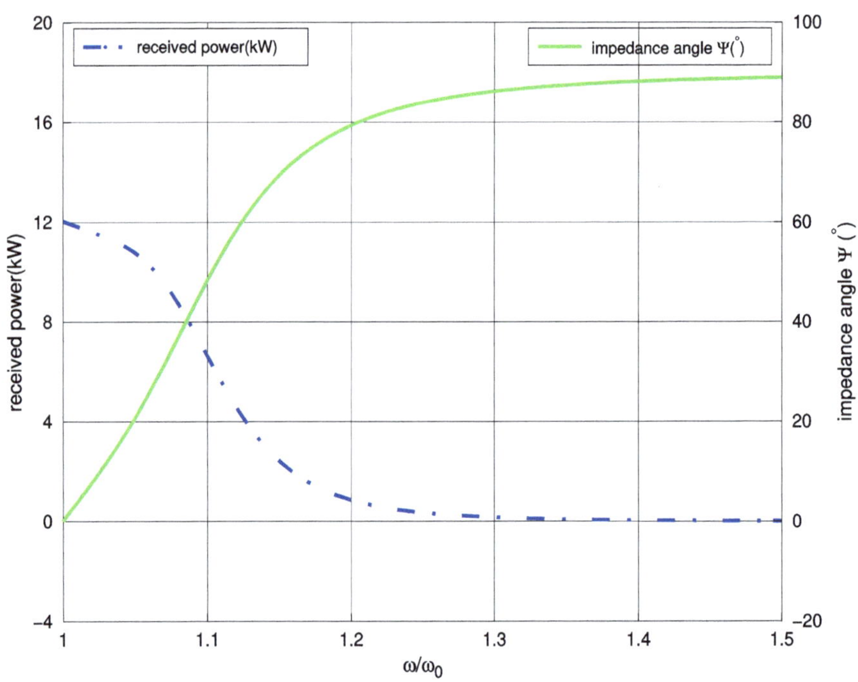

Fig. 2.2 Received power and impedance angle versus ω/ω_0 under the parameters listed in Table 2.1

Table 2.1 Parameters of a WPT system for simulation

Symbol	Parameter	Value	Symbol	Parameter	Value
L	Resonant coil inductance	33 μH	r_s	On-resistance of MOSFET	0.05 Ω
C	Resonant capacitance	106.24 nF	R_L	Load resistance	5 Ω
k	Coupling coefficient	0.2	V_I	DC bus voltage	400 V
r	ESR of coil and capacitor	0.05 Ω	f_0	Self-resonant frequency	85 kHz

ESR r is assumed constant for the simulation. When r is much smaller than the load resistance, as in Table 2.1, the assumption is acceptable [1, 3, 4]. In order to obtain zero-voltage switching (ZVS) for the inverter, the operating frequency should be higher than the self-resonant frequency [5], namely, ψ should be higher than 0. Hence, only operating frequencies above the self-resonant frequency are employed for this simulation.

It can be seen that when the operating frequency increases, the received power decreases. Hence, frequency adjustment can be employed as a method to regulate the transferred power in some cases. The input impedance angle ψ is higher than zero except at $\omega/\omega_0 = 1$.

In practice, the self-resonant frequencies of both sides are hardly the same. Figure 2.3 plots the received power and impedance angle versus ω/ω_0 under two cases, $\omega_2/\omega_0 = 0.95$ and $\omega_2/\omega_0 = 1.05$, where ω_0 and ω_2 are the self-resonant angular frequencies of the sending side and the receiving side, respectively. Note that ω_0 for the receiving side in Eq. (2.22) should be changed to ω_2 for this simulation. The remaining parameters are the same as those listed in Table 2.1, except the resonant capacitor at the receiving side, whose values are 117.7 nF and 96.4 nF, respectively, to obtain the corresponding resonant frequencies. Figure 2.3a shows that the input impedance angle is below zero when $\omega/\omega_0 < 1.033$ for the given case $\omega_2/\omega_0 = 0.95$, which means the inverter must operate at $\omega/\omega_0 > 1.033$ to obtain a ZVS stage.

Besides the operating frequency, the coupling coefficient k strongly influences the transferred power because it varies rapidly with the distance between the two coils. For example, $k \approx 0.2$ for a gap of 20 cm for a 4-turn coil with widths 90 cm × 70 cm [3], while $k \approx 0.088$ at 100 cm for a 21-turn coil with 50 cm diameters. Figure 2.4 depicts the transferred power and impedance angle versus the coupling coefficient under the parameters listed in Table 2.1, except for the coupling coefficient itself.

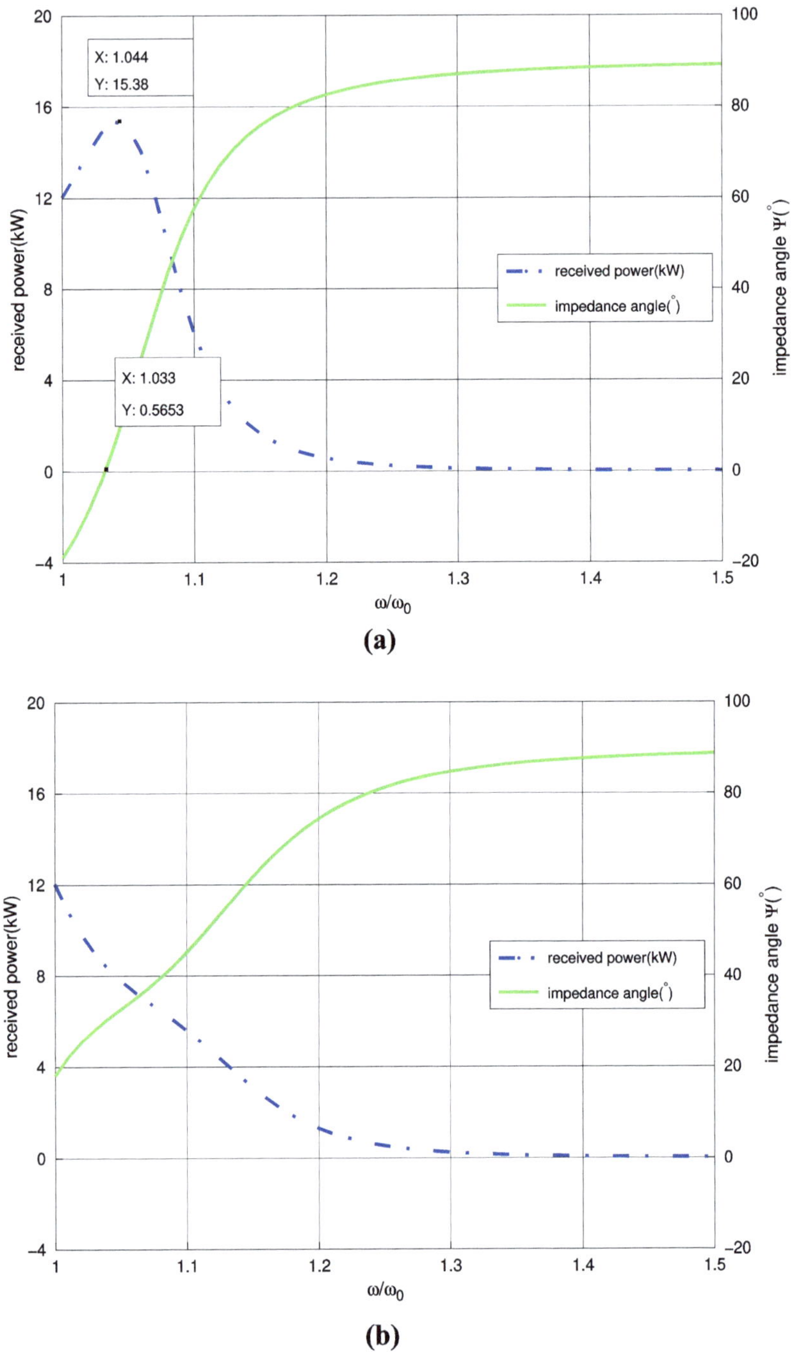

Fig. 2.3 Received power and impedance angle versus ω/ω_0 under different self-resonant frequencies at the two sides, where (**a**) $\omega_2/\omega_0 = 0.95$ and (**b**) $\omega_2/\omega_0 = 1.05$

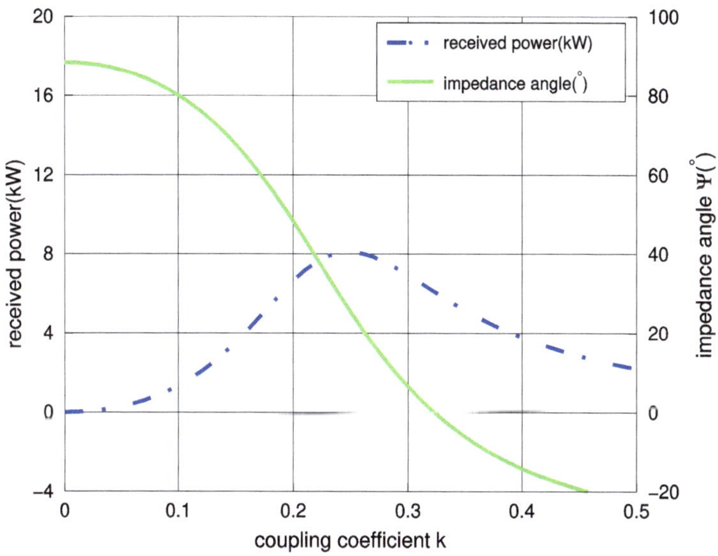

Fig. 2.4 Received power and impedance angle versus k under $\omega/\omega_0 = 1.1$

Efficiency is another important consideration besides the transferred power. When the switching loss of the inverter is ignored, the efficiency is

$$\eta = \frac{P_2}{P_i} = \frac{P_2}{V_{\mathrm{rms}}\, I_{1,\mathrm{rms}} \cos\psi} = \frac{\left(\dfrac{0.45\,\omega M V_I}{|Z_2|\,|Z_1 + Z_{\mathrm{ref}}|}\right)^2 R_L}{\left(\dfrac{0.45\,V_I}{|Z_1 + Z_{\mathrm{ref}}|}\right) 0.45\,V_I \cos\psi}$$
$$= \frac{\left(\dfrac{0.45\,\omega M V_I}{|Z_2|\,|Z_1 + Z_{\mathrm{ref}}|}\right)^2 R_L}{\left(\dfrac{0.45\,V_I}{|Z_1 + Z_{\mathrm{ref}}|}\right) 0.45\,V_I \left(\dfrac{\mathrm{real}(Z_1 + Z_{\mathrm{ref}})}{|Z_1 + Z_{\mathrm{ref}}|}\right)} = \frac{(\omega M)^2 R_L}{\mathrm{real}(Z_1 + Z_{\mathrm{ref}})\,|Z_2|^2} \tag{2.23}$$

Additionally, the efficiency can be expressed as the product of two terms, respectively at the sending and receiving sides,

$$\eta = \eta_1 \times \eta_2 = \frac{P_{\mathrm{trans}}\ \mathrm{real}(Z_{\mathrm{ref}})}{P_{\mathrm{in}}\ \mathrm{real}(Z_1 + Z_{\mathrm{ref}})} \times \frac{P_2\, R_L}{P_{\mathrm{trans}}\ \mathrm{real}(Z_2)}$$
$$= \frac{\mathrm{real}(Z_{\mathrm{ref}})}{\mathrm{real}(Z_1 + Z_{\mathrm{ref}})} \times \frac{R_L}{\mathrm{real}(Z_2)} \tag{2.24}$$

where real(·) denotes the real part. The two expressions in Eqs. (2.23) and (2.24) obtain the same result, because Eq. (2.24) can be transformed into

$$\eta = \frac{\text{real}(Z_{\text{ref}})}{\text{real}(Z_1 + Z_{\text{ref}})} \times \frac{R_L}{\text{real}(Z_2)} = \frac{\text{real}\left(\frac{(\omega M)^2}{Z_2}\right)}{\text{real}(Z_1 + Z_{\text{ref}})} \times \frac{R_L}{\text{real}(Z_2)}$$
$$= \frac{(\omega M)^2 R_L \,\text{real}\left(\frac{1}{Z_2}\right)}{\text{real}(Z_1 + Z_{\text{ref}})\,\text{real}(Z_2)} = \frac{(\omega M)^2 R_L}{\text{real}(Z_1 + Z_{\text{ref}})\,|Z_2|^2} \tag{2.25}$$

Substitution of Eqs. (2.20) and (2.21) into Eq. (2.23) yields, with the shorthand $\Phi := Q_0\left(\frac{\omega}{\omega_0} - \frac{\omega_0}{\omega}\right)$,

$$\eta = \frac{(\omega M)^2 R_L}{\text{real}\left(r(1 + j\Phi) + r_s + \frac{(\omega M)^2}{r(1+j\Phi)+R_L}\right) \times |r(1 + j\Phi) + R_L|^2}$$
$$= \frac{(\omega M)^2 R_L}{\left(r + r_s + \frac{(r + R_L)(\omega M)^2}{(r + R_L)^2 + r\Phi^2}\right) \times \left((r + R_L)^2 + r\Phi^2\right)}$$
$$= \frac{(\omega k L)^2 R_L}{(r + r_s)\left((r + R_L)^2 + r\Phi^2\right) + (r + R_L)(\omega k L)^2} \tag{2.26}$$

Figure 2.5 depicts the efficiency versus ω/ω_0 under the parameters listed in Table 2.1, except for k. One can see that the highest efficiency occurs at a frequency close to the self-resonant frequency for various coupling coefficients k, and the efficiency decreases as the operating frequency shifts away from the self-resonant frequency. Additionally, a higher k provides a higher efficiency at a given frequency.

The parameters listed in Table 2.1 correspond to a typical electrical-bus wireless charger [3], where the gap between the two coils is relatively small and yields a high coupling coefficient, while the operating frequency is relatively low, which results in a parasitic resistance of 50 mΩ. However, for WPT systems with large gaps, the operating frequency is commonly a few hundreds of kilohertz or even a few megahertz, and the coil parasitic resistance may rise to several ohms. As a result, the on-state resistance of the switches and the losses in the capacitors can be ignored compared with the coil parasitic resistance. Assuming the system operates under the resonant condition

$$\eta = \frac{(\omega k L)^2 R_L}{r(r + R_L)^2 + (r + R_L)(\omega k L)^2} = \frac{\left(\frac{\omega k L}{r}\right)^2}{1 + \left(\frac{\omega k L}{r}\right)^2 + \frac{R_L}{r}} \cdot \frac{R_L/r}{1 + R_L/r} \tag{2.27}$$

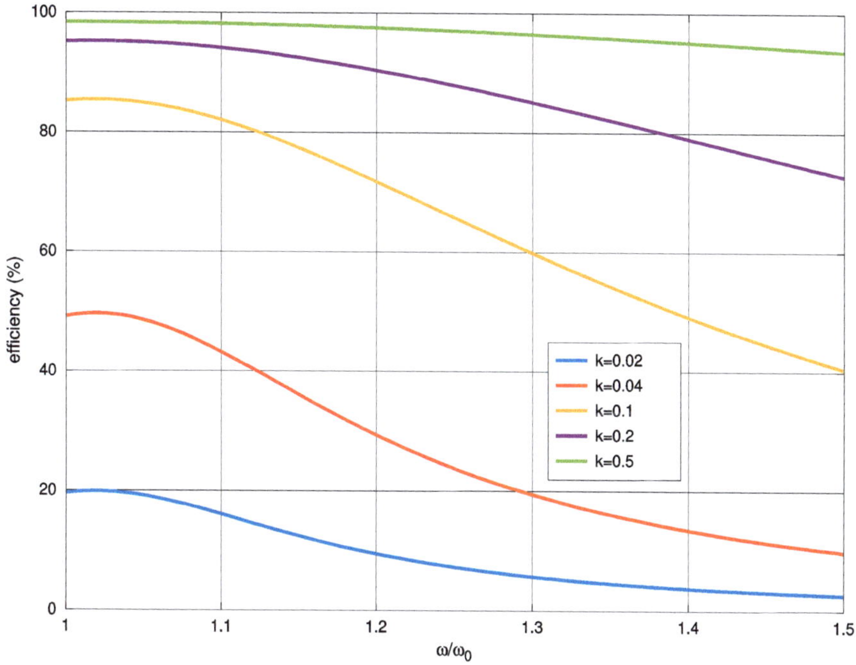

Fig. 2.5 Efficiency versus ω/ω_0 under the parameters listed in Table 2.1, except for k

the loaded quality factor is defined as

$$Q_L = \frac{\omega L}{R_L} \tag{2.28}$$

Substitution of Eqs. (2.21) and (2.28) into Eq. (2.27) yields

$$\eta = \frac{k^2}{k^2 + \dfrac{1}{Q_0 Q_L} + \dfrac{1}{Q_0^2}} \cdot \frac{1/Q_L}{1/Q_0 + 1/Q_L} \tag{2.29}$$

The highest efficiency is achieved under the condition $\partial\eta/\partial Q_L = 0$, which yields the optimized loaded quality factor $Q_{L,\mathrm{opt}}$ and the highest efficiency $\eta_{\max}$ as expressed in Eqs. (2.30) and (2.31), respectively:

$$Q_{L,\mathrm{opt}} = \frac{Q_0}{\sqrt{1 + k^2 Q_0^2}} \tag{2.30}$$

$$\eta_{\max} = \left(\frac{kQ_0}{1 + \sqrt{1 + k^2 Q_0^2}} \right)^2 \tag{2.31}$$

It can be seen from Eq. (2.31) that a higher unloaded quality factor Q_0 results in a higher $\eta_{\max}$. Under the optimized loaded quality factor $Q_{L,\text{opt}}$, Eq. (2.28) is rewritten as

$$Q_{L,\text{opt}} = \frac{\omega_0 L}{R_{L\text{opt}}} \tag{2.32}$$

which leads to the optimized load resistance

$$R_{L,\text{opt}} = \frac{\omega_0 L}{Q_{L,\text{opt}}} = \frac{\omega_0 L}{\frac{Q_0}{\sqrt{1 + (kQ_0)^2}}} = r\sqrt{1 + (kQ_0)^2} \tag{2.33}$$

This leads to an important conclusion: a higher efficiency can always be achieved under a higher unloaded quality factor Q_0 by adjusting the load resistance according to Eq. (2.33). Under the parameters listed in Table 2.1, the optimized load resistance is

$$R_{L,\text{opt}} = \sqrt{1 + 0.2^2 \times 352.28^2} \times 0.05 = 3.5252\,\Omega \tag{2.34}$$

and the maximum efficiency under the optimized load resistance, from Eq. (2.31), is

$$\eta_{\max} = \left(\frac{0.2 \times 352.28}{1 + \sqrt{1 + (0.2 \times 352.28)^2}} \right)^2 = 0.972 \tag{2.35}$$

2.1.3 Design Examples

Example 2.1 Design a wireless power transfer system employing the topology shown in Fig. 2.1a. The WPT system meets the following specifications: the input impedance angle is $\psi = 20°$, and the remaining parameters are listed in Table 2.1. Neglecting inverter switching losses, evaluate the operating frequency ω, the received power P_2, and the efficiency η.
Solution: Based on Eq. (2.11), one obtains

$$\psi = \arctan\left(\frac{\text{imag}(Z_1 + Z_{\text{ref}})}{\text{real}(Z_1 + Z_{\text{ref}})}\right)$$

$$= \arctan\left(\frac{\omega L\left(\frac{\omega}{\omega_0} - \frac{\omega_0}{\omega}\right) - \frac{(\omega k L)^2\,\omega L\left(\frac{\omega}{\omega_0} - \frac{\omega_0}{\omega}\right)}{(r+R_L)^2 + \left(\omega L\left(\frac{\omega}{\omega_0} - \frac{\omega_0}{\omega}\right)\right)^2}}{r + r_s + \frac{(\omega k L)^2 (r+R_L)}{(r+R_L)^2 + \left(\omega L\left(\frac{\omega}{\omega_0} - \frac{\omega_0}{\omega}\right)\right)^2}}\right) = \frac{20\pi}{180} \tag{2.36}$$

Substituting the values listed in Table 2.1 into Eq. (2.36) yields a function with the variable ω. Using MATLAB's `solve` function, one obtains the operating angular frequency, namely $f = 89.1\,\text{kHz}$.

Using Eqs. (2.4), (2.5), and (2.8), the impedances at the sending loop, the receiving loop, and the reflected branch are

$$\begin{aligned} Z_1 &= j\omega L + \frac{1}{j\omega C_1} + r + r_s \\ &= j(560.01 \times 10^3)(33 \times 10^{-6}) + 0.05 + 0.05 \\ &\quad + \frac{1}{j(560.01 \times 10^3)(106.24 \times 10^{-9})} \\ &= 0.10 + j1.6723 \end{aligned} \tag{2.37}$$

$$\begin{aligned} Z_2 &= j\omega L + \frac{1}{j\omega C_2} + r + R_L \\ &= j(560.01 \times 10^3)(33 \times 10^{-6}) + 0.05 + 5 \\ &\quad + \frac{1}{j(560.01 \times 10^3)(106.24 \times 10^{-9})} \\ &= 5.05 + j1.6723 \end{aligned} \tag{2.38}$$

$$\begin{aligned} Z_{\text{ref}} &= \frac{(\omega M)^2}{Z_2} \\ &= \frac{\left((560.01 \times 10^3)(33 \times 10^{-6} \times 0.2)\right)^2}{5.05 + j1.6723} \\ &= 2.438 - j0.807 \end{aligned} \tag{2.39}$$

Substitution of Eqs. (2.37)–(2.39) into Eq. (2.18) yields

$$
\begin{aligned}
P_2 &= \left(\frac{0.45\omega M V_I}{|Z_2|\,|Z_1 + Z_{\text{ref}}|}\right)^2 R_L \\
&= \left(\frac{0.45(560.01 \times 10^3)(6.6 \times 10^{-6})(400)}{|5.05 + j1.6723|\,|0.10 + j1.6723 + 2.438 - j0.807|}\right)^2 \times 5 \\
&= 10.87\,\text{kW}
\end{aligned}
\tag{2.40}
$$

Using Eq. (2.24), the efficiency is

$$
\begin{aligned}
\eta &= \frac{\text{real}(Z_{\text{ref}})}{\text{real}(Z_1 + Z_{\text{ref}})} \cdot \frac{R_L}{\text{real}(Z_2)} \\
&= \frac{\text{real}(2.438 - j0.807)}{\text{real}(0.10 + j1.6723 + 2.438 - j0.807)} \cdot \frac{5}{\text{real}(5.05 + j1.6723)} \\
&= 95.11\%
\end{aligned}
\tag{2.41}
$$

Example 2.2 If a full-bridge inverter is employed instead of the half-bridge inverter in Example 2.1 for the same WPT system, the operating angular frequency is $\omega = 560.01\,\text{krad/s}$, while the other parameters are the same as those listed in Table 2.1. Neglecting inverter switching losses, evaluate the input impedance angle, the received power P_2, and the efficiency η.
Solution: Using Eqs. (2.4), (2.5), and (2.8), and considering that the ESR of the switch is twice that of the half-bridge case, the impedances at the sending loop, the receiving loop, and the reflected branch are

$$
\begin{aligned}
Z_1 &= j\omega L + \frac{1}{j\omega C_1} + r_1 + 2r_s \\
&= j(560.01 \times 10^3)(33 \times 10^{-6}) + 0.05 + 2 \times 0.05 \\
&\quad + \frac{1}{j(560.01 \times 10^3)(106.24 \times 10^{-9})} \\
&= 0.15 + j1.6723
\end{aligned}
\tag{2.42}
$$

$$
\begin{aligned}
Z_2 &= j\omega L + \frac{1}{j\omega C_2} + r_2 + R_L \\
&= j(560.01 \times 10^3)(33 \times 10^{-6}) + 0.05 + 5 \\
&\quad + \frac{1}{j(560.01 \times 10^3)(106.24 \times 10^{-9})} \\
&= 5.05 + j1.6723
\end{aligned}
\tag{2.43}
$$

Solution: According to Eq. (2.15), the rms value of the fundamental component current through the sending loop is

$$\begin{aligned} I_{1,\text{rms}} &= \frac{V_{\text{rms}}}{|Z_1 + Z_{\text{ref}}|} \approx \frac{0.45 V_I}{|Z_1 + Z_{\text{ref}}|} \\ &= \frac{0.45 \times 400}{|0.10 + j1.6723 + 2.438 - j0.807|} \\ &= 67.11\,\text{A} \end{aligned} \tag{2.48}$$

Since the switches operate under ZVS, the turn-on loss is zero and only the turn-off loss is considered. The turn-off loss for an inverter arm is [5]

$$\begin{aligned} P_{SSW} &= \frac{\omega \left(\frac{t_r}{3} + \frac{t_f}{2}\right) V_I \sqrt{2} I_{1,\text{rms}} \sin \psi}{2\pi} \\ &= \frac{(560.01 \times 10^3)\left(\frac{28}{3} + \frac{8}{2}\right) \times 10^{-9} \times 400 \times (\sqrt{2} \times 67.11) \sin 20^\circ}{2\pi} \\ &= 15.43\,\text{W} \end{aligned} \tag{2.49}$$

where t_r and t_f are the rising time and falling time of the MOSFET, respectively. According to Eq. (2.24), and including the switching loss, the efficiency is

$$\begin{aligned} \eta = \eta_1 \eta_2 &= \frac{I_{1,\text{rms}}^2 \,\text{real}(Z_{\text{ref}})}{I_{1,\text{rms}}^2 \,\text{real}(Z_1 + Z_{\text{ref}}) + P_{SSW}} \cdot \frac{I_{2,\text{rms}}^2 R_L}{I_{2,\text{rms}}^2 \,\text{real}(Z_2)} \\ &= \frac{67.11^2 \,\text{real}(2.438 - j0.807)}{67.11^2 \,\text{real}(0.10 + j1.6723 + 2.438 - j0.807) + 15.43} \\ &\quad \cdot \frac{R_L}{\text{real}(5.05 + j1.6723)} \\ &= 94.98\% \end{aligned} \tag{2.50}$$

Comparison of Examples 2.1 and 2.3 shows that the switching loss has a very small influence on the overall efficiency.

2.2 Other Compensation Schemes

Besides the Series-Series (S-S) compensation, Parallel-Series (P-S), Series-Parallel (S-P) and Parallel-Parallel (P-P) compensations are three other schemes which can be applied in wireless power transfer. These are discussed below.

2.2.1 Parallel-series Compensation

The equivalent circuit of the Parallel-Series compensation wireless power transfer supplied by a voltage source is shown in Fig. 2.6. The parallel-series compensation is achieved by using capacitance C_1 connected in parallel with the primary inductance L_1; on the secondary side, capacitance C_2 is connected in series to both the load resistance R_L and secondary inductance L_2.

Furthermore, the analysis of the WPT topology based on Fig. 2.6 is carried out in the frequency domain. Due to the increase in the complexity of topologies, the equivalent impedance method no longer simplifies analysis. We introduce here basic loop analysis and nodal analysis methods instead.

One obtains the impedances of two parts of the sending side (Z_1, Z_2) and the receiving side (Z_S) as

$$Z_1 = j\omega L_1 + R_1 \tag{2.51}$$

$$Z_2 = \frac{1}{j\omega C_1} \tag{2.52}$$

$$Z_S = j\omega L_2 + \frac{1}{j\omega C_2} + R_2 + R_L \tag{2.53}$$

Applying KVL to both the sending and receiving sides, the core equation set is obtained,

$$\begin{cases} v' = Z_P i_{L_1} + j\omega M i_2 \\ 0 = j\omega M i_{L_1} + Z_S i_2 \end{cases} \tag{2.54}$$

where

$$Z_P = Z_1 + Z_2 \parallel r_s = Z_1 + \frac{Z_2 r_s}{Z_2 + r_s} \tag{2.55}$$

$$V' = \frac{Z_2 V}{Z_2 + r_s} \tag{2.56}$$

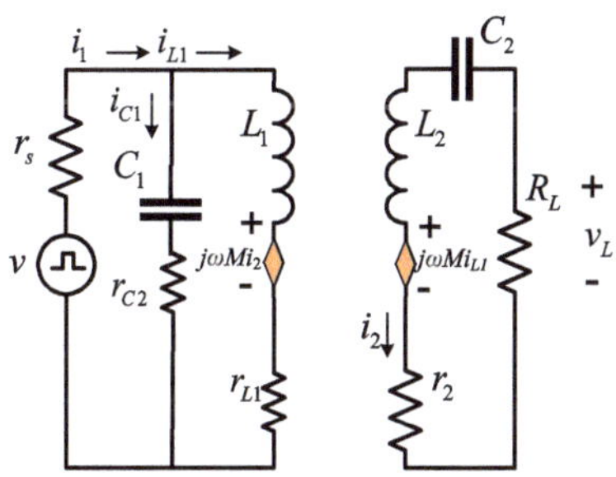

Fig. 2.6 Equivalent circuits of P-S model

From (2.54) one obtains the current through the respective inductance of the sending loop and receiving loop as

$$I_{L_1} = \frac{V' Z_S}{Z_P Z_S + (\omega M)^2} \tag{2.57}$$

Using the rms relation for the fundamental component,

$$V_{\text{rms}} \approx 0.45\, V, \qquad V'_{\text{rms}} = \frac{0.45\, V\, Z_2}{Z_2 + r_s} \tag{2.58}$$

Then, based on the KVL and KCL laws in the frequency domain at the sending side, an equation set is obtained:

$$\begin{cases} I_1 = I_{L_1} + I_{C_1} \\ V - r_s I_1 = Z_2 I_{C_1} = Z_1 I_{L_1} + j\omega M I_2 \end{cases} \tag{2.59}$$

Solving it, an expression for I_1 in terms of I_{L_1} is found:

$$I_1 = \frac{V + Z_2\, I_{L_1}}{Z_2 + r_s} \tag{2.60}$$

In addition, since the overall circuit impedance is difficult to compute, the impedance angle ψ is defined as

$$\psi = \psi_V - \psi_{I_1} = 0 - \psi_{I_1} = -\arctan\left(\frac{\text{imag}(I_1)}{\text{real}(I_1)}\right) \tag{2.61}$$

Next, based on (2.57) and (2.58), the rms value of the fundamental component current through the inductance of the sending loop is

$$I_{L_1,\text{rms}} = \frac{V'_{\text{rms}}\, Z_S}{\left|Z_P Z_S + (\omega M)^2\right|} \tag{2.62}$$

Hence, one obtains the rms value of the fundamental component current through the inductance of the receiving loop as

$$I_{L_2,\text{rms}} = \frac{\omega M\, I_{L_1,\text{rms}}}{|Z_S|} = \frac{\omega M\, V'_{\text{rms}}}{\left|Z_P Z_S + (\omega M)^2\right|} \tag{2.63}$$

The transferred power and efficiency will be calculated next. It is assumed that the circuit is loaded with a pure resistance R_L. The rms value of the voltage across the load R_L and power received are

$$V_{L,\mathrm{rms}} = I_{L_2,\mathrm{rms}}\, R_L = \frac{\omega M\, V'_{\mathrm{rms}}\, R_L}{\left|Z_P Z_S + (\omega M)^2\right|} \tag{2.64}$$

$$P_{2\text{-PS}} = I_{L_2,\mathrm{rms}}\, V_{L,\mathrm{rms}} = \frac{(\omega M\, V'_{\mathrm{rms}})^2\, R_L}{\left|Z_P Z_S + (\omega M)^2\right|^2} \tag{2.65}$$

Substituting (2.53), (2.55) and (2.58) into (2.65), one obtains the transferred power depending on the operating frequency as

$$P_{2\text{-PS}} = \left| \frac{0.45\,\omega M V \cdot \dfrac{1}{1 + j\omega C_1 R_s}}{\left(j\omega L_2 + \frac{1}{j\omega C_2} + R_2 + R_L\right)\left(j\omega L_1 + R_1 + \frac{R_s}{1+j\omega C_1 R_s}\right) + (\omega M)^2} \right|^2 R_L \tag{2.66}$$

2.2.1.1 Compensation Design Method

In order to reduce the reactive power in the circuit and improve transmission efficiency, designing the compensation capacitors on the sending and receiving side plays an important role in the overall design. The principle of compensation design is to cancel the imaginary part of the circuit, as discussed below.

The capacitance C_2 in the receiving side is chosen to resonate with the secondary coil at the resonance frequency $f_0 = \omega_0/(2\pi)$, thus

$$C_2 = \frac{1}{\omega_0^2\, L_2} \tag{2.67}$$

Using the method of reflecting resistance, the impedance seen by the source V is

$$Z_{IN} = \frac{1}{j\omega C_1 + \dfrac{1}{R_1 + j\omega L_1 + \frac{(\omega M)^2}{R_2 + R_L}}} \cdot \frac{1}{j\omega C_1 r_s + 1} \tag{2.68}$$

Neglecting the parasitic resistance of the primary and secondary coils R_1 and R_2 and on-resistance of MOSFET r_s, the value of C_1 that cancels the imaginary part of the input impedance Z_{IN} at the resonant frequency ω_0 is [6, 7]

$$C_1 = \frac{L_1}{\omega_0^2 \left(L_1^2 + \dfrac{(\omega_0 M)^4}{R_L^2}\right)} \tag{2.69}$$

From (2.69) we can find that in parallel-series compensation, besides the inductance L_1 and resonant frequency ω_0, the primary capacitance also depends on the coupling coefficient k of primary and secondary inductance and the load resistance R_L.

Additionally, from the above we can find that the reflected resistance is a pure resistance when the secondary side achieves resonant state, which has no impact on the imaginary part of the primary side. That is to say, we can design the compensation capacitance in each side respectively in such topology.

2.2.1.2 Parametric Simulation

According to the compensation design method and the given parameters, the parameters of P-S compensation model for simulation are listed in Table 2.2 as follows.

As is shown in Fig. 2.7, when the operating frequency increases, the received power does not vary monotonously and it obtains maximum power around the self-resonant frequency. Hence, frequency adjusting can be employed as a method to regulate the transferred power in some cases.

Besides the operating frequency, the coupling coefficient k plays a significant influence on the transferred power. It can be seen in Fig. 2.8 that the transferred power versus the coupling coefficient under parameters listed in Table 2.2 except the coupling coefficient.

The input impedance angle ψ is higher than zero when f is below the self-resonant frequency; to satisfy the ZVS angle, 74.4 kHz is chosen to be the operating frequency.

The efficiency is another concerned issue besides the transferred power. The efficiency is defined by the power received by R_L divided by that theoretical output

Table 2.2 Parameters of P-S compensation model

Symbol	Parameter explanation	Value
V	DC bus voltage	350 V
f_0	Self-resonant frequency	80 kHz
k	Coefficient of mutual inductance	0.2
L_1	Resonant inductance on sending side	1 μH
L_2	Resonant inductance on receiving side	34 μH
C_1	Designed capacitance on sending side	3.78 μF
C_2	Designed capacitance on receiving side	116 nF
R_1	ESR of coil on sending side	0.04 Ω
R_2	ESR of coil on receiving side	0.04 Ω
r_s	On-resistance of MOSFET	0.04 Ω
R_L	Resistance of the load	5 Ω

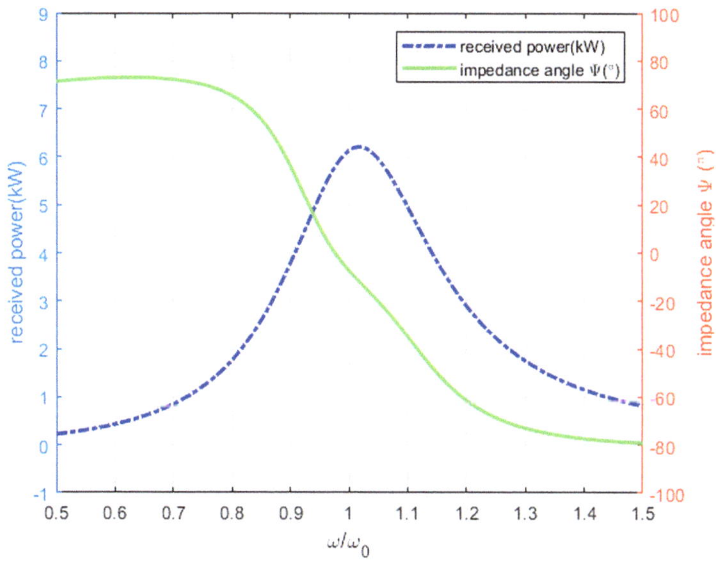

Fig. 2.7 Received power and impedance angle vs. f under parameters listed in Table 2.2

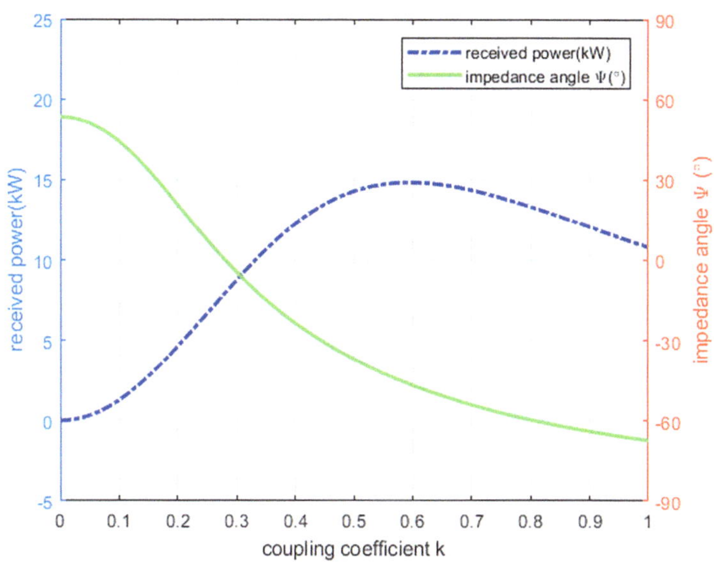

Fig. 2.8 Received power and impedance angle vs. k under operating frequency (74.4 kHz)

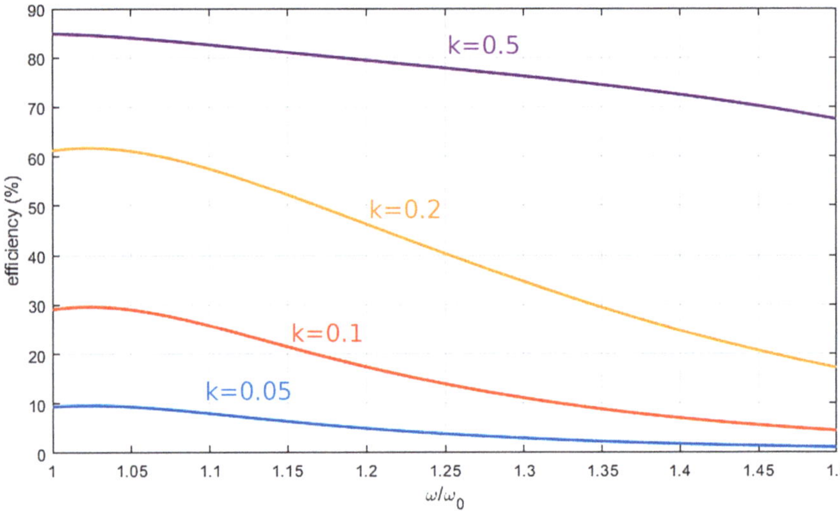

Fig. 2.9 Efficiency vs. f under parameters listed in Table 2.2 except k

from the inverter. When the switching loss of inverters is ignored, from (2.55) the efficiency is

$$\eta_{\text{PS}} = \frac{P_2}{P_{\text{in}}} = \frac{P_2}{\text{Re}(I_1) \times V_{\text{rms}}} \tag{2.70}$$

as shown in Fig. 2.9.

2.2.2 Series-parallel Compensation

The equivalent circuit of series-parallel compensation is achieved by using capacitance C_1 connected in series both to the primary inductance L_1 and voltage source V. On the secondary side, capacitance C_2 is connected in parallel with the load resistance R_L.

Following, the analysis of the WPT topology based on Fig. 2.10 is carried out in the frequency domain. One obtains the impedance at the sending side (Z_P) and two parts of the receiving side (Z_3, Z_4) as

$$Z_P = j\omega L_1 + \frac{1}{j\omega C_1} + R_1 + r_s \tag{2.71}$$

$$Z_3 = j\omega L_2 + R_2 \tag{2.72}$$

$$Z_4 = \frac{1}{j\omega C_2} \tag{2.73}$$

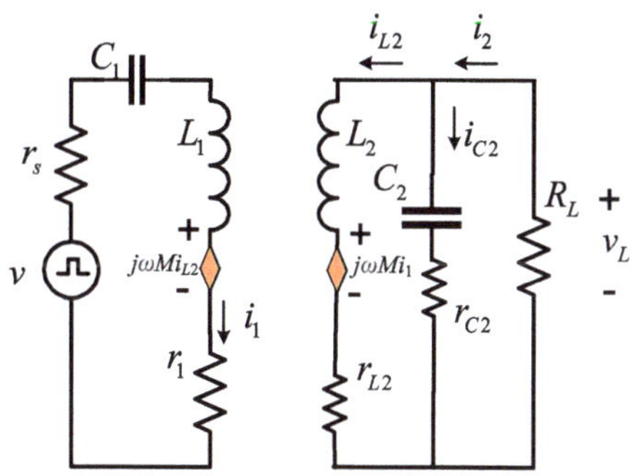

Fig. 2.10 Equivalent circuits of S-P model

Based on the KVL and KCL laws in the frequency domain at the sending side and receiving side, the governing equations

$$\begin{cases} V = Z_P I_1 + j\omega M I_{L_2} \\ 0 = j\omega M I_1 + Z_S I_{L_2} \end{cases} \tag{2.74}$$

is obtained, where

$$Z_S = Z_3 + Z_4 \parallel R_L = Z_3 + \frac{Z_4 R_L}{Z_4 + R_L} \tag{2.75}$$

(and (2.58) also applies here).

From now, the deducing progress of the S-P topology will be the same as the P-S topology; thus the expressions obtained in the last section can be used directly by replacing Z_S. From (2.63)–(2.65):

$$I_{1,\mathrm{rms}} = I_{L_1,\mathrm{rms}} = \frac{V_{\mathrm{rms}} Z_S}{\left|Z_P Z_S + (\omega M)^2\right|} \tag{2.76}$$

$$I_{2rms} = I_{L2rms} \times \left|\frac{Z_4}{Z_4 + R_L}\right| = \frac{\omega M V_{rms}}{\left|Z_P Z_S + (\omega M)^2\right|} \times \left|\frac{Z_4}{Z_4 + R_L}\right| \tag{2.77}$$

$$P_{2-SP} = I_{2rms}^2 R_L = \left(\frac{\omega M V_{rms} \times \left|\frac{Z_4}{Z_4+R_L}\right|}{\left|Z_P Z_S + (\omega M)^2\right|}\right)^2 R_L \tag{2.78}$$

Substituting (2.71)–(2.73) into (2.78), the power of the load is obtained as

$$P_{2-SP} = \left(\frac{0.45\omega M V \times \left| \frac{1}{1+j\omega C_2 R_L} \right|}{|D_{\text{SP}}|} \right)^2 R_L,$$

$$D_{\text{SP}} = \left(j\omega L_2 + R_2 + \frac{R_L}{1+j\omega C_2 R_L} \right) \cdot \left(j\omega L_1 + \frac{1}{j\omega C_1} + R_1 + r_s \right) + (\omega M)^2 \tag{2.79}$$

2.2.2.1 Compensation Design Method

The secondary capacitance C_2 is chosen to resonate with the secondary coil at the resonance frequency, thus [6, 7]

$$C_2 = \frac{1}{\omega_0^2 L_2} \tag{2.80}$$

The impedance seen from the source V is

$$Z_{\text{IN}} = R_1 + j\omega L_1 - \frac{j}{\omega C_1} + \frac{(\omega M)^2}{R_2 + j\omega L_2 + \frac{R_L}{1 + j\omega C_2 R_L}} \tag{2.81}$$

Neglecting the parasitic resistance of the primary and secondary coils R_1, R_2, the value of C_1 that cancels the imaginary part of the input impedance at the resonant frequency ω_0 is

$$C_1 = \frac{L_2}{\omega_0^2 (L_1 L_2 - M^2)} \tag{2.82}$$

In the series-parallel compensation, besides the resonant frequency ω_0 and primary inductance, the primary capacitance C_1 also depends on the coupling coefficient k of the primary and secondary inductance L_2. However, it has nothing to do with the load resistance R_L.

2.2.2.2 Parametric Simulation

According to the compensation design method and the given parameters, the parameters of S-P compensation model for simulation are listed in Table 2.3 as follows.

As is shown in Fig. 2.11, when the operating frequency increases, the received power obtains maximum power around the self-resonant frequency. Hence, fre-

Table 2.3 Parameters of S-P compensation model

Symbol	Parameter explanation	Value
V	DC bus voltage	350 V
f_s	Self-resonant frequency	85 kHz
k	Coefficient of mutual inductance	0.2
L_1	Resonant inductance on sending side	33.6 μH
L_2	Resonant inductance on receiving side	33.6 μH
C_1	Resonant capacitance on sending side	108.7 nF
C_2	Resonant capacitance on receiving side	104.3 nF
R_1	ESR of coil on sending side	0.04 Ω
R_2	ESR of coil on receiving side	0.04 Ω
r_s	On-resistance of MOSFET	0.04 Ω
R_L	Resistance of the load	50 Ω

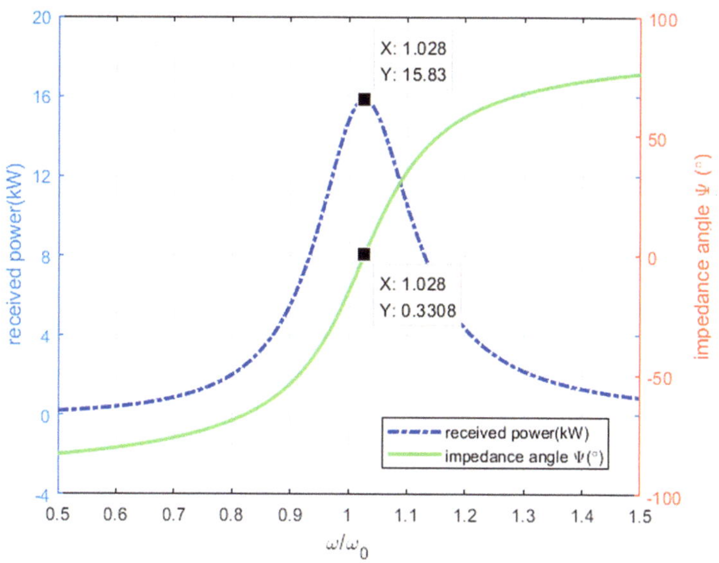

Fig. 2.11 Received power and impedance angle vs. f under parameters listed in Table 2.3

quency adjusting can be employed as a method to regulate the transferred power in some cases.

Besides the operating frequency, the coupling coefficient k plays an important role in the transferred power. As can be seen from Fig. 2.12, the transferred power versus the coupling coefficient under parameters listed in Table 2.3 except the coupling coefficient.

The input impedance angle ψ is higher than zero when f is above the self-resonant frequency; to satisfy the ZVS angle, 87.2 kHz is chosen to be the operating frequency. From Fig. 2.12 we can find that receiving power does not perform in direct proportion to coupling coefficient k but achieves maximum when k is around 0.2.

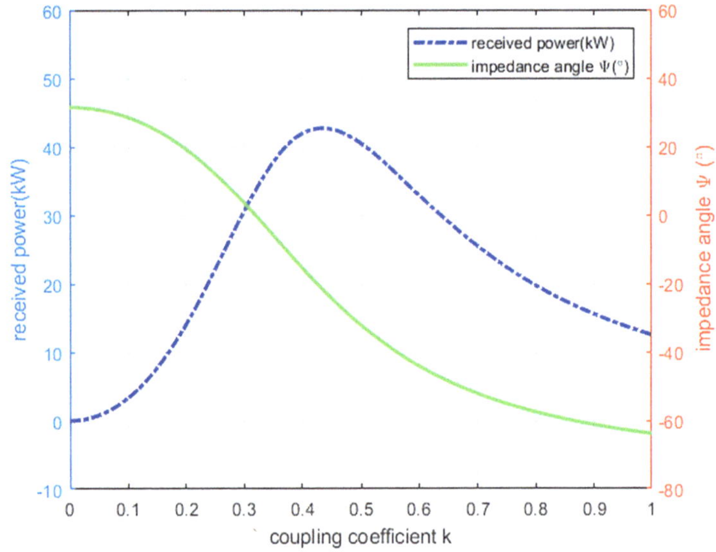

Fig. 2.12 Received power and impedance angle vs. k under operating frequency (87.2 kHz)

Efficiency is another concerned issue besides the transferred power. The efficiency is defined by the power received by R_L divided by that theoretical output from the inverter. When the switching loss of inverters is ignored, from (2.55) the efficiency is

$$\eta_{SP} = \frac{P_2}{P_1} = \frac{P_2}{\mathrm{Re}(I_1) \times V_{\mathrm{rms}}} \tag{2.83}$$

as shown in Fig. 2.13.

2.2.3 *Parallel-parallel Compensation*

The equivalent circuits of the Parallel-Parallel compensation wireless power transfer supplied by a voltage source is shown in Fig. 2.14.

Then, the analysis of the WPT based on Fig. 2.14 is carried out in the frequency domain. Similarly, based on the KVL and KCL laws in the frequency domain at the sending side and receiving side, the core equation set is obtained:

$$\begin{cases} V' = Z_P I_{L_1} + j\omega M I_2 \\ 0 = j\omega M I_{L_1} + Z_S I_2 \end{cases} \tag{2.84}$$

where (2.55), (2.75) and (2.56) apply.

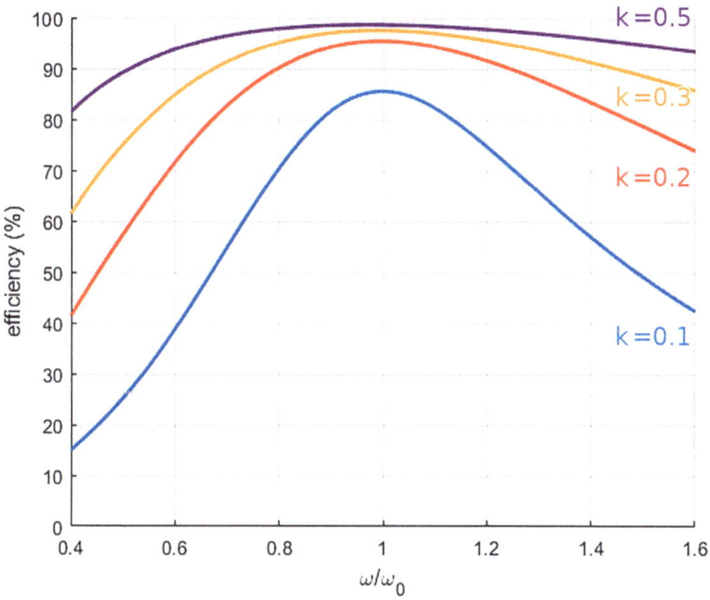

Fig. 2.13 Efficiency vs. f under parameters listed in Table 2.3 except k

Fig. 2.14 Equivalent circuits of P-P model

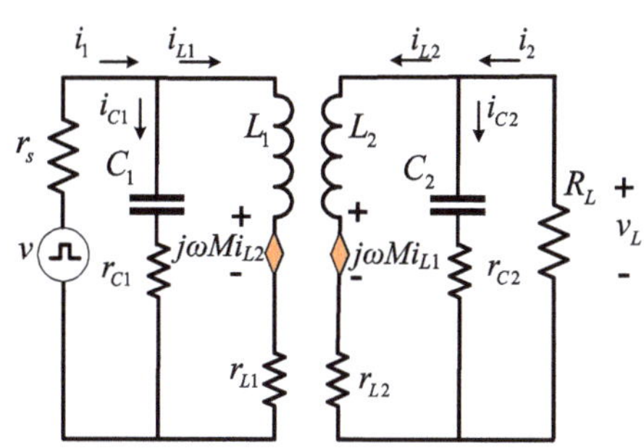

Given the preceding derivation, the equations of the P-P topology will be the same as those of the former sections; thus the expressions obtained previously can be used directly by replacing Z_S. Thus

$$P_{2\text{-PP}} = I_{2rms}^2 R_L = \left| \frac{\omega M \, V_{\text{rms}}}{Z_P Z_S + (\omega M)^2} \cdot \frac{Z_4}{Z_4 + R_L} \cdot \frac{Z_2}{Z_2 + r_s} \right|^2 R_L \tag{2.85}$$

Substituting (2.55), (2.75) and (2.72) into (2.85), the received power is obtained as

$$P_{2\text{-PP}} = \left(\frac{0.45\,\omega M\,V \cdot \left| \dfrac{Z_4}{Z_4 + R_L} \right| \cdot \left| \dfrac{Z_2}{Z_2 + r_s} \right|}{\left| Z_P Z_S + (\omega M)^2 \right|} \right)^2 R_L$$

$$= \left(\frac{0.45\omega M V \times \left| \frac{1}{1+j\omega C_1 r_s} \right| \times \left| \frac{1}{1+j\omega C_2 R_L} \right|}{|D_{\text{PP}}|} \right)^2 R_L,$$

$$D_{\text{PP}} = \left(j\omega L_2 + R_2 + \frac{R_L}{1+j\omega C_2 R_L} \right) \cdot \left(j\omega L_1 + R_1 + \frac{r_s}{1+j\omega C_1 r_s} \right) + (\omega M)^2 \tag{2.86}$$

2.2.3.1 Compensation Design Method

The secondary capacitance C_2 is chosen to resonate with the secondary coil at the resonance frequency, thus

$$C_2 = \frac{1}{\omega_0^2 L_2} \tag{2.87}$$

The impedance seen from the source V is

$$Z_{IN} = (r_s + \frac{1}{j\omega C_1}) \parallel \left(Z_P + \frac{(\omega M)^2}{Z_S} \right) \tag{2.88}$$

Neglecting the parasitic resistance of the primary and secondary coils R_1, R_2 and on-resistance of MOSFET r_s,

$$Z_{\text{IN}} = \frac{1}{j\omega C_1 + \dfrac{1}{j\omega L_1 + \dfrac{(\omega M)^2}{j\omega L_2 + \dfrac{1}{j\omega C_2}}}} \tag{2.89}$$

Thus the value of C_1 that cancels the imaginary part of the input impedance at the resonant frequency ω_0 is

$$C_1 = \frac{L_2^3 (L_1 L_2 - M^2)}{M^4 R_L^2 + \omega_0^2 L_2^2 (L_1 L_2 - M^2)^2} \tag{2.90}$$

Table 2.4 Parameters of P-P compensation model

Symbol	Parameter explanation	Value
V	DC bus voltage	350 V
f_s	Self-resonant frequency	80 kHz
k	Coefficient of mutual inductance	0.2
L_1	Resonant inductance on sending side	2.5 μH
L_2	Resonant inductance on receiving side	2.5 μH
C_1	Designed capacitance on sending side	1.60 μF
C_2	Designed capacitance on receiving side	1.58 μF
R_1	ESR of coil on sending side	0.04 Ω
R_2	ESR of coil on receiving side	0.04 Ω
r_s	On-resistance of MOSFET	0.04 Ω
R_L	Resistance of the load	5 Ω

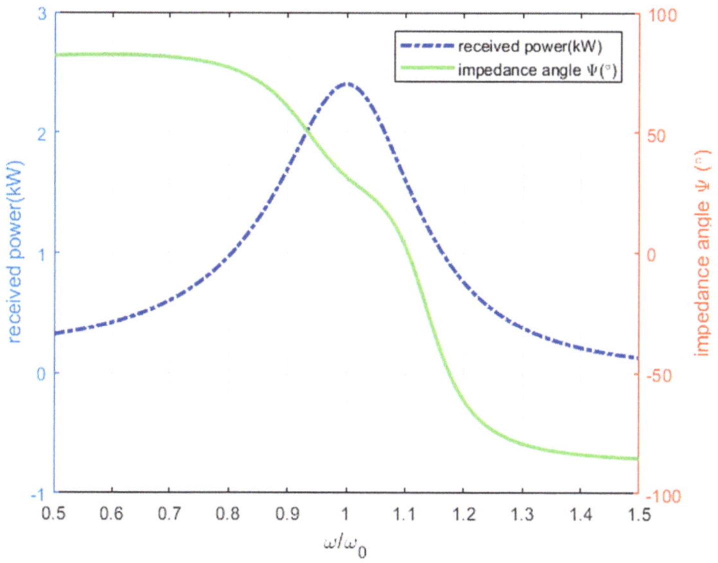

Fig. 2.15 Received power and impedance angle vs. f under parameters listed in Table 2.4

In the parallel-parallel compensation, the design of C_1 is complicated and depends on many parameters of the circuit.

2.2.3.2 Parametric Simulation

According to the compensation design method and the given parameters, the parameters of P-P compensation model for simulation are listed in Table 2.4 as follows. Fig. 2.15 depicts the received power and impedances angle vs. frequency.

It can be seen that the received power obtains maximum power around the self-resonant frequency. Therefore, frequency adjusting can be employed as a method

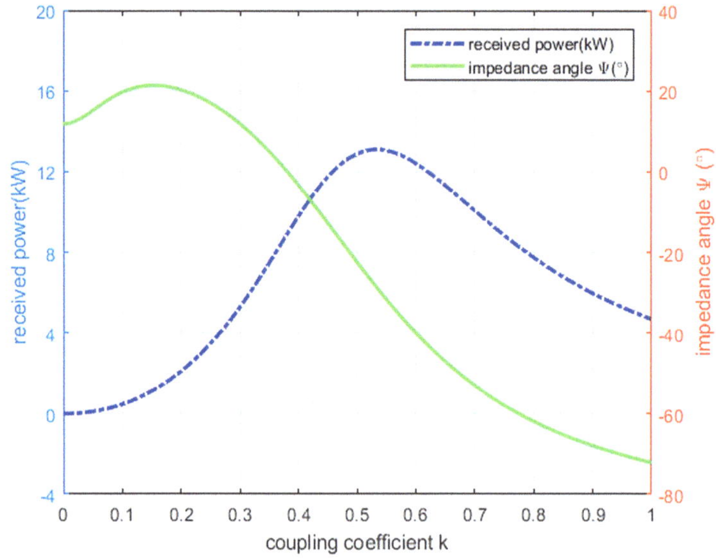

Fig. 2.16 Received power and impedance angle vs. k under operating frequency (75.6 kHz)

to regulate the transferred power in some cases. The input impedance angle ψ is higher than zero when f is below the self-resonant frequency.

Besides the operating frequency, the coupling coefficient k is vital in adjusting the transferred power. It can be seen in Fig. 2.16 that the transferred power versus the coupling coefficient under parameters listed in Table 2.4 except the coupling coefficient.

The efficiency is another concerned issue besides the transferred power. The efficiency is defined by the power received by R_L divided by the theoretical output from the inverter. When the switching loss of inverters is ignored, the efficiency is

$$\eta_{\mathrm{PS}} = \frac{P_2}{P_{\mathrm{in}}} = \frac{P_2}{\mathrm{Re}(I_1) \times V_{\mathrm{rms}}} \tag{2.91}$$

as shown in Fig. 2.17.

2.2.4 *Comparisons Between Different Topologies*

In the last three sections, we briefly introduced four basic compensation topologies' circuit configurations, analysis procedures and computing methods. We also need to learn about each topology's specialties for practical use. Based on theoretical analysis and experience conclusions from other references, the strengths and weaknesses of each compensation topology are concluded in Table 2.5 [6–9].

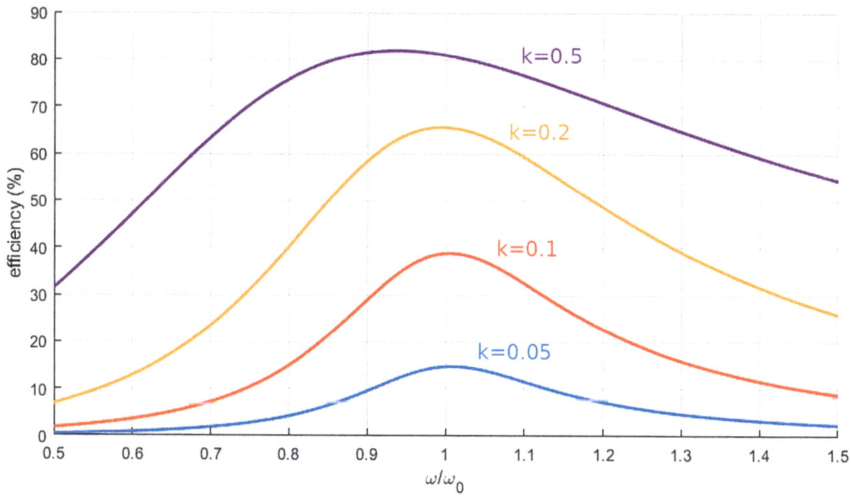

Fig. 2.17 Efficiency vs. f under parameters listed in Table 2.4 except k

Table 2.5 Comparisons of different compensation models

Topology	Strengths	Weaknesses
S-S	Simple in structure and easy to design.	Source has no protection. Transmission efficiency decreases with higher load resistance.
S-P	Transmission efficiency increases with higher load resistance.	Source has no protection. Higher voltage stress on reactive power devices of sending side compared with S-S.
P-S	Source has protective capacity. Lower voltage stress on reactive power devices of sending side compared with S-S.	Higher current stress on capacity of sending side. Transmission efficiency decreases with higher load resistance.
P-P	Source has protective capacity.	Transmission efficiency is very low at any load resistance. Complicated to analyze and difficult to design. The value of designed capacitance is very high, thus it is impractical.

Example 2.4 Design a wireless power transfer system for an appliance with large load resistance. The parameters of the WPT system specifications are shown in Table 2.6. Without considering the inverter switching loss, design the value of each capacitor on the sending side and receiving side C_1, C_2 and evaluate the output power P_2 and transmission efficiency η.

The first step of designing a WPT system is to choose an appropriate compensation topology. From [6, 7], the transfer efficiency of the S-P topology increases with the increase of load resistance, which is suitable to be applied in a system with high load resistance. Hence the system design is based on Fig. 2.10.

Table 2.6 Known parameters of a WPT system (Example 2.4)

Symbol	Parameter	Value	Symbol	Parameter	Value
L_1, L_2	Resonant coil inductance	33.6 μH	r_s	On-resistance of MOSFET	0.04 Ω
k	Coupling coefficient	0.2	R_L	Load resistance	50 Ω
R_1, R_2	ESR of coil	0.04 Ω	V_I	DC bus voltage	350 V
f	Self-resonant frequency	85 kHz	f_o	Operating frequency	88 kHz

From Table 2.6, the self-resonant frequency is 85 kHz ($\omega = 534071$ rad/s). In order to make the sending and receiving side achieve self-resonance, each loop's imaginary part should be zero.

For the receiving side:

$$C_2 = \frac{1}{\omega^2 L_2} = \frac{1}{(2\pi f)^2 L_2} = \frac{1}{(2\pi \times 85 \times 10^3)^2 \times 33.6 \times 10^{-6}} = 104.34\,\text{nF}$$

For the sending side, the impedance seen by V is

$$Z_{\text{IN}} = R_1 + j\omega L_1 - \frac{j}{\omega C_1} + \frac{\omega^2 M^2}{R_2 + j\omega L_2 + \dfrac{R_L}{1 + j\omega C_2 R_L}}$$

To make C_1 cancel the imaginary part of Z_{IN}:

$$C_1 = \frac{L_2}{\omega_0^2 (L_1 L_2 - M^2)}$$
$$= \frac{33.6 \times 10^{-6}}{(5.34 \times 10^5)^2 \times \left[(33.6 \times 10^{-6})^2 - (6.72 \times 10^{-6})^2\right]} = 108.7\,\text{nF}$$

All the circuit component parameters are now determined and listed in Table 2.7.

Table 2.7 Designed parameters of a WPT system (Example 2.4)

Symbol	Parameter	Value	Symbol	Parameter	Value
L_1, L_2	Resonant coil inductance	33.6 μH	r_s	On-resistance of MOSFET	0.04 Ω
k	Coupling coefficient	0.2	R_L	Load resistance	50 Ω
C_1	Sending-side capacitance	108.7 nF	C_2	Receiving-side capacitance	104.34 nF
f	Self-resonant frequency	85 kHz	f_o	Operating frequency	88 kHz
R_1, R_2	ESR of coil	0.04 Ω	r_{C_1}, r_{C_2}	ESR of capacitor	0.01 Ω

Based on (2.75):

$$P_2 = \left(\frac{\omega M V_{\text{rms}} \cdot \left| \frac{Z_4}{Z_4 + R_L} \right|}{\left| Z_P Z_S + (\omega M)^2 \right|} \right)^2 R_L$$

$$= \left(\frac{0.45 \times 534 \times 10^3 \times 6.72 \times 10^{-6} \times 350 \times |0.11 - 0.32j|}{|(5.74 + 2.04j)(0.08 + 0.71j) + 12.88|} \right)^2 \times 50$$

$$= 11.49\,\text{kW}$$

Based on (2.76):

$$\eta = \frac{P_2}{\text{Re}(I_1) \times V_{\text{rms}}} = \frac{P_2}{\text{Re}\left(\frac{0.45V\,Z_S}{Z_S Z_P + (\omega M)^2} \right) \times 0.45V}$$

$$= \frac{11.49 \times 10^3}{76.01 \times 0.45 \times 350} = 95.99\%$$

2.3 Multi-element Resonant Inverters

Multi-element resonant inverters contain multiple reactive components in the resonant tank, such as LCC, LCC–LCC, CLL wireless power transfer (WPT) systems, and so on, which are depicted in Fig. 2.18. In this section, we take the LCC WPT system as an example. A combination of parallel compensation on the primary side with series compensation on the secondary side is widely used because of its

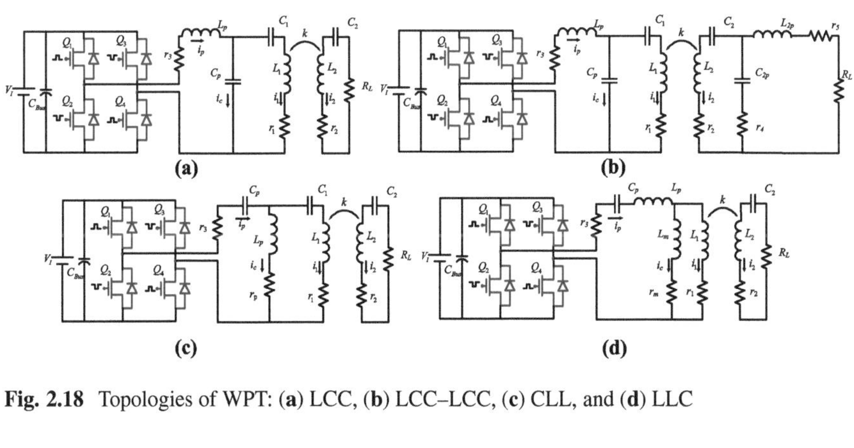

Fig. 2.18 Topologies of WPT: (**a**) LCC, (**b**) LCC–LCC, (**c**) CLL, and (**d**) LLC

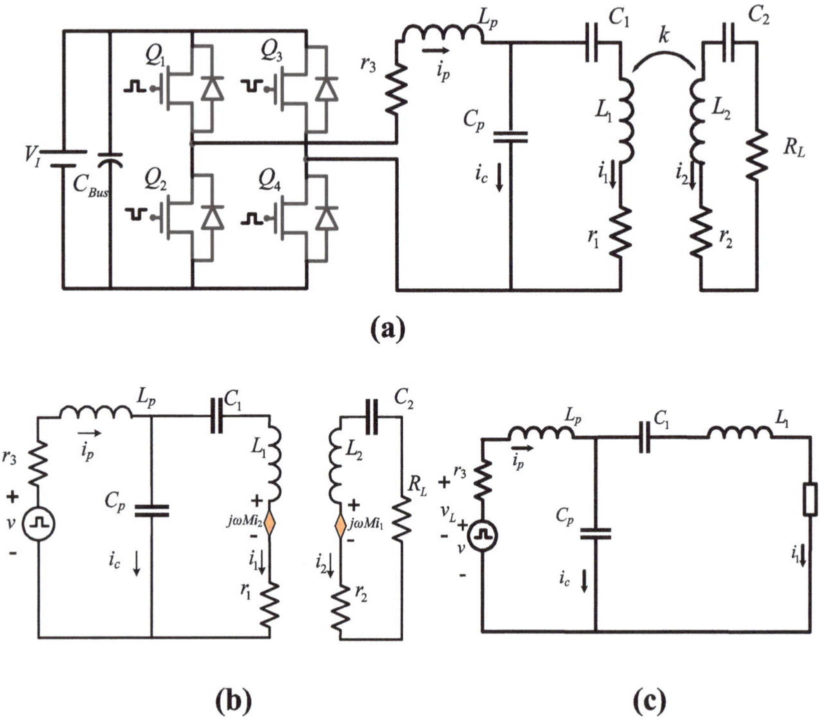

Fig. 2.19 Electrical model of a wireless power transfer system employing an LCC compensation topology: (**a**) full bridge, and (**b**)–(**c**) equivalent circuits

flexibility and safety. The full-bridge multi-element resonant inverter is detailed in the following [10–12].

2.3.1 Topologies

Figure 2.19a shows the complete power-stage topology of LCC-C WPT, which consists of a full-bridge (FB) inverter, an LCC compensator on the primary side, coils, and a series compensator capacitor on the secondary side with the FB rectifier. The inverter output is replaced by a square wave in Fig. 2.19b, and the influence of the coupling inductors is replaced by two current-dependent voltage sources in Fig. 2.19c [10].

It can be seen in Fig. 2.19 that the LCC-C compensation topology includes multiple components, such as a series resonant inductor L_p, a parallel resonant capacitor C_p, a series capacitor in the transmitter C_1, and a series capacitor on the receiver side C_2. The coupling factor between the two coils is expressed by

$$M = k\sqrt{L_1 L_2} \tag{2.92}$$

where M is the mutual inductance between L_1 and L_2.

From Eq. (2.12), the voltage can be expressed as

$$v = V_m \sin \omega t \tag{2.93}$$

where V_m is the amplitude of the fundamental component per

$$V_m = \frac{4V}{\pi} \tag{2.94}$$

Thus, the rms value of the voltage is

$$V_{\text{rms}} = \frac{V_m}{\sqrt{2}} \approx 0.9V \tag{2.95}$$

Equation (2.96) expresses the impedance of the receiver side Z_S, which can be reflected to the transmitter side as Z_{ref} as shown in Eq. (2.97). Therefore, the circuit can be simplified as in Fig. 2.19c.

$$Z_S = j\omega L_2 + \frac{1}{j\omega C_2} + R_L + r_2 \tag{2.96}$$

$$Z_{\text{ref}} = \frac{\omega^2 M^2}{Z_S} \tag{2.97}$$

We assume that the fundamental resonant frequency ω_0 is given by

$$\omega_0 = \frac{1}{\sqrt{L_p C_p}} \tag{2.98}$$

2.3.2 Characteristics

Furthermore, the analysis of the LCC-compensation WPT topology based on Fig. 2.19b is carried out in the frequency domain. Here we introduce this topology using the basic loop and nodal analysis methods.

One obtains the impedances at three parts of the sending side (Z_1, Z_2, Z_3) and the receiving side as

$$Z_1 = j\omega L_1 + r_1 + \frac{1}{j\omega C_1} \tag{2.99}$$

$$Z_2 = \frac{1}{j\omega C_p} \tag{2.100}$$

$$Z_3 = j\omega L_p + r_3 \tag{2.101}$$

$$Z_S = j\omega L_2 + r_2 + \frac{1}{j\omega C_2} + R_L \tag{2.102}$$

Based on the KVL and KCL laws in the frequency domain at the sending side, an equation set is obtained:

$$\begin{cases} i_p = i_c + i_1, \\ v - i_p(j\omega L_p + r_3) = Z_2 i_c = Z_1 i_1 + j\omega M i_2. \end{cases} \tag{2.103}$$

The i_p and i_1 can be expressed as

$$i_p = \frac{v + Z_2 i_1}{Z_2 + Z_3} \tag{2.104}$$

Substitution of Eqs. (2.99)–(2.101) into Eq. (2.103), together with the use of KVL at the receiving side, leads to the core equation set

$$\begin{cases} v' = Z_p i_1 + j\omega M i_2, \\ 0 = j\omega M i_1 + Z_S i_2. \end{cases} \tag{2.105}$$

where

$$Z_p = Z_1 + Z_2 \parallel Z_3 = Z_1 + \frac{Z_2 Z_3}{Z_2 + Z_3} \tag{2.106}$$

$$v' = \frac{Z_2}{Z_2 + Z_3} \cdot v \tag{2.107}$$

From Eqs. (2.106) and (2.107), one obtains the currents i_1 and i_2 as

$$i_1 = \frac{v' Z_S}{Z_p Z_S + (\omega M)^2} \tag{2.108}$$

$$i_2 = -\frac{j\omega M i_1}{Z_S} = -\frac{j\omega M v'}{Z_p Z_S + (\omega M)^2} \tag{2.109}$$

From Eq. (2.107), V_{rms} can be expressed as

$$V_{\text{rms}} \approx 0.9V, \qquad V'_{\text{rms}} = \frac{0.9V Z_2}{Z_2 + Z_3} \tag{2.110}$$

Based on Eqs. (2.108) and (2.110), the rms value of the fundamental component current through the inductance of the sending loop is

$$I_{1,\text{rms}} \approx \frac{V'_{\text{rms}} Z_S}{\left|Z_p Z_S + (\omega M)^2\right|} \tag{2.111}$$

Therefore, one obtains the rms value of the fundamental component current through the inductance of the receiving loop as

$$I_{2,\text{rms}} \approx \frac{\omega M V'_{\text{rms}}}{\left|Z_p Z_S + (\omega M)^2\right|} \tag{2.112}$$

It is assumed that the circuit is loaded with a pure resistance R_L. The rms value of the voltage across the load and the received power are

$$V_{L,\text{rms}} \approx I_{2,\text{rms}} R_L = \frac{\omega M V'_{\text{rms}} R_L}{\left|Z_p Z_S + (\omega M)^2\right|} \tag{2.113}$$

$$P_{2\text{-}LCC} = I_{2,\text{rms}} V_{L,\text{rms}} = \left(\frac{\omega M V'_{\text{rms}}}{\left|Z_p Z_S + (\omega M)^2\right|}\right)^2 R_L \tag{2.114}$$

One obtains the transferred power depending on the operating frequency as. For brevity, let $Z_c := \frac{1}{j\omega C_p} + r_1$:

$$P_{2\text{-}LCC} = \left|\frac{0.9\,\omega M V \times \left|\frac{Z_c}{Z_c + Z_3}\right|}{\left(j\omega L_2 + \frac{1}{j\omega C_2} + r_2 + R_L\right)\cdot\left(j\omega L_1 + r_1 + \frac{Z_c \cdot Z_3}{Z_c + Z_3}\right) + (\omega M)^2}\right|^2 R_L \tag{2.115}$$

Since the whole impedance of the circuit is difficult to calculate, the impedance angle ψ is defined as

$$\psi = -\arctan\frac{\text{imag}(i_p)}{\text{real}(i_p)} \tag{2.116}$$

The parameters of the LCC compensation model are listed in Table 2.8.

Table 2.8 Parameters of LCC-C compensation model for transferred-power and impedance-angle plots

Symbol	Parameter explanation	Value
V	DC bus voltage	80 V
f	Operating frequency	20 kHz
L_p	Inductance connected with the inverter	123 μH
k	Coefficient of mutual inductance	0.2
L_1	Resonant inductance on sending side	550.6 μH
L_2	Resonant inductance on receiving side	334.8 μH
C_1	Resonant capacitance on sending side	192.7 nF
C_2	Resonant capacitance on receiving side	196.3 nF
r_1	ESR of coil on sending side	0.3 Ω
r_3	ESR of coil of L_p	0.11 Ω
r_2	ESR of coil and capacitance on receiving side	0.17 Ω
C_p	Resonant compensation capacitance	504.7 nF
R_L	Resistance of the load	2.5 Ω

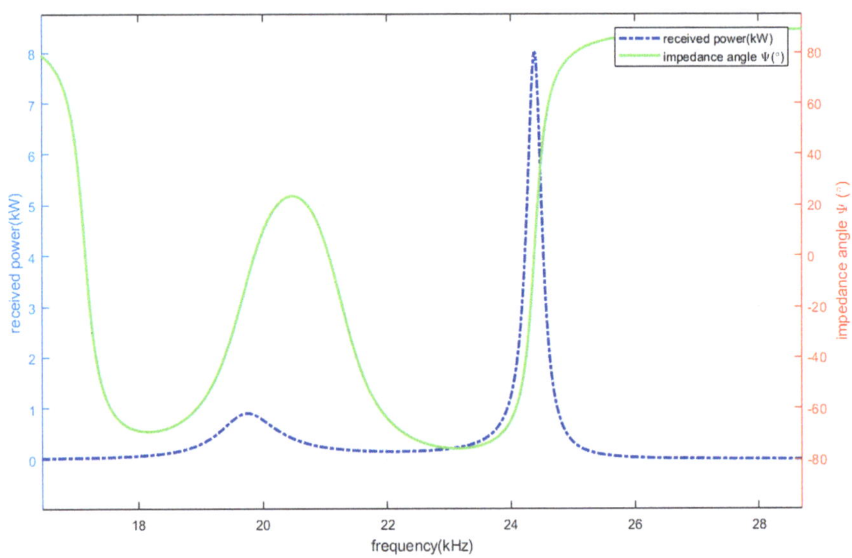

Fig. 2.20 Received power and impedance angle vs. operating frequency under parameters listed in Table 2.8

As shown in Fig. 2.20, when the frequency increases, the received power does not vary monotonically and reaches a maximum around 24 kHz. Therefore, the transferred power can be regulated by adjusting the frequency.

Besides the operating frequency, the coupling coefficient k plays an important role in the transferred power. Figure 2.21 shows that the input impedance angle ψ is positive.

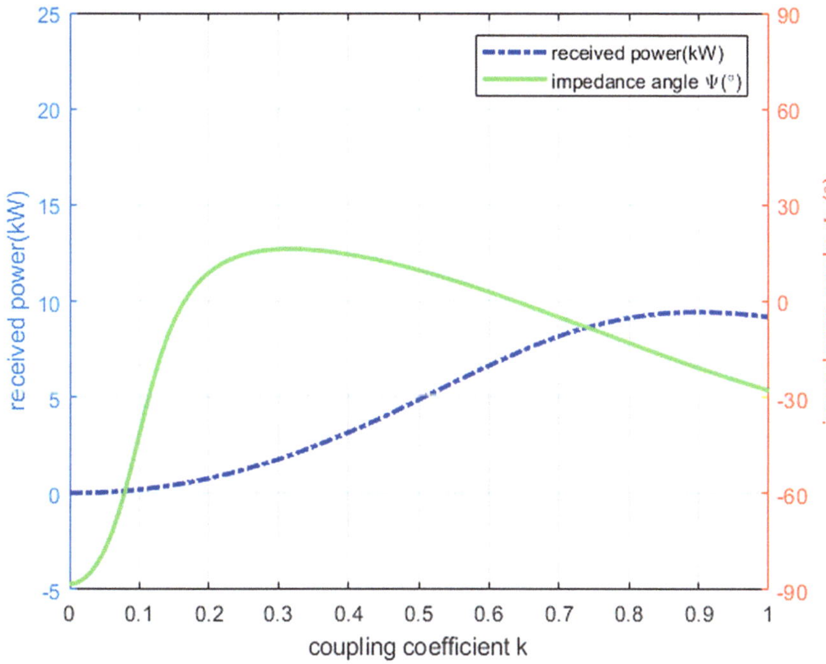

Fig. 2.21 Received power and impedance angle vs. coupling coefficient under operating frequency (24.6 kHz)

To satisfy the ZVS angle, 24.6 kHz is chosen to be the operating frequency.

2.3.2.1 Voltage-gain Characteristic

Based on KVL, v_i can be expressed by Eq. (2.117). Meanwhile, setting the operating frequency to ω_0 eliminates the impedance of L_p and C_p, so i_1 can be written as Eq. (2.120).

$$v_i = i_p j\omega L_p + i_C\left(\frac{1}{j\omega C_p}\right); \quad i_C = i_p - i_1 \tag{2.117}$$

$$v_i = i_p\left(j\omega L_p + \frac{1}{j\omega C_p}\right) - i_1\left(\frac{1}{j\omega C_p}\right) \tag{2.118}$$

$$j\omega L_p + \frac{1}{j\omega C_p} = 0 \tag{2.119}$$

$$i_1 = \omega C_p v_i \tag{2.120}$$

Table 2.9 Parameters of LCC-C compensation model for voltage-gain plots

Symbol	Parameter explanation	Value
f_0	Operating frequency	20 kHz
L_p	Inductance connected with the inverter	30 μH
k	Coefficient of mutual inductance	0.2
L_1	Resonant inductance on sending side	160 μH
L_2	Resonant inductance on receiving side	160 μH
C_1	Resonant capacitance on sending side	14.22 nF
C_2	Resonant capacitance on receiving side	14.22 nF
C_p	Resonant compensation capacitance	34.7 nF

It can be seen from Eq. (2.120) that the primary-coil current i_1 is independent of the load resistance and is constant as long as the inverter output voltage and switching frequency do not vary, which is an important characteristic of LCC-C WPT.

The total impedance of WPT Z_T and the output-to-input transfer function G_v are

$$Z_T = j\omega L_p + \frac{\frac{1}{j\omega C_p}\left(\frac{1}{j\omega C_1} + j\omega L_1 + Z_{\text{ref}}\right)}{\frac{1}{j\omega C_p} + \frac{1}{j\omega C_1} + j\omega L_1 + Z_{\text{ref}}} \tag{2.121}$$

$$G_v = \frac{V_o}{V_{\text{in}}} = \frac{j\omega M R_L \left(1 - \omega^2 L_p C_p - j\omega C_p Z_T\right)}{Z_T Z_S} \tag{2.122}$$

The parameters of the LCC compensation model used for the voltage-gain plots are listed in Table 2.9.

Figure 2.22 shows a load-independent frequency. Owing to the existence of multiple reactive components, however, the load-independent frequency is not ω_0.

In addition, it is useful to investigate the behavior of the WPT system under different coupling coefficients k, as shown in Fig. 2.23.

2.3.3 Design Example

Design a wireless power transfer system for an appliance with 2.5 Ω resistance. The parameters of the WPT system are shown in Table 2.10. Without considering the inverter switching loss, design the values of the capacitances on the sending and receiving sides and choose the appropriate frequency to obtain the desired output power and efficiency.

In order to make the sending and receiving sides achieve self-resonance, each loop's imaginary part should be zero.

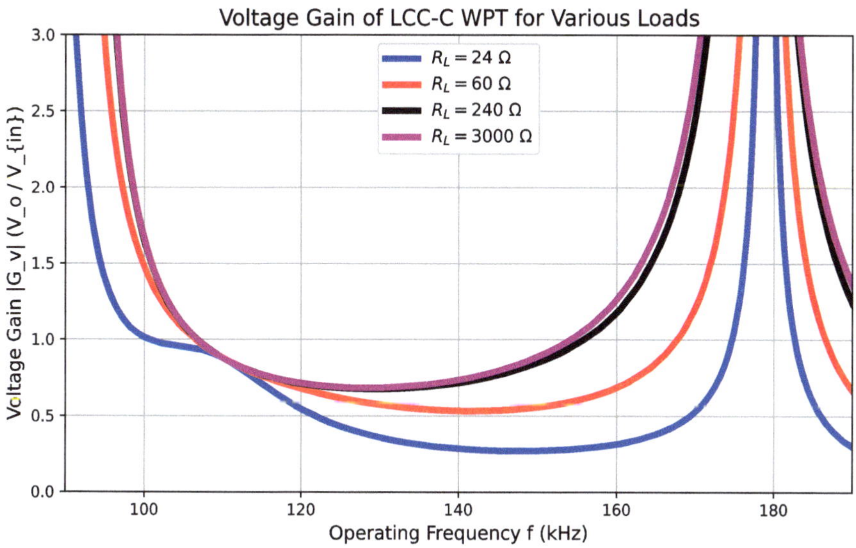

Fig. 2.22 Voltage gain of LCC-C WPT for various loads

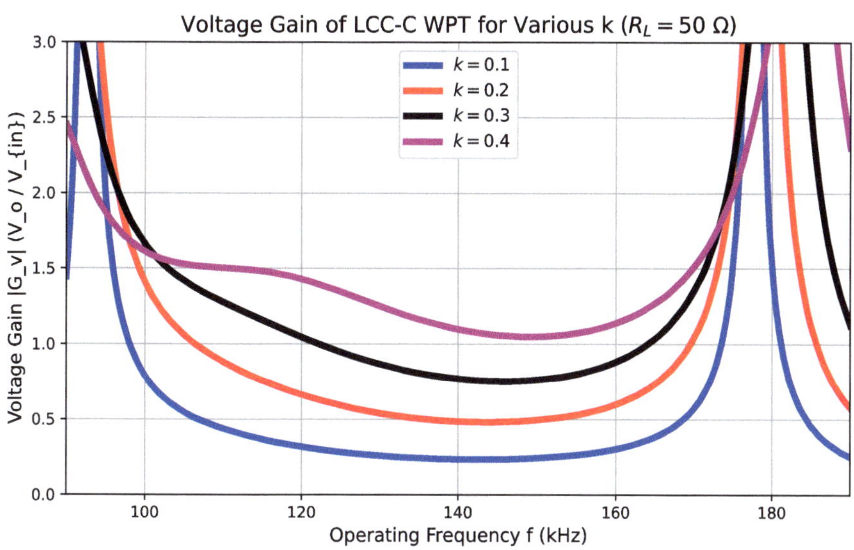

Fig. 2.23 Voltage gain of LCC-C WPT for various coupling coefficients

Table 2.10 Known parameters of the WPT system used in the design example

Symbol	Parameter	Value
L_1	Resonant inductance on sending side	550.6 μH
L_2	Resonant inductance on receiving side	334.8 μH
k	Coupling coefficient	0.2
R_L	Load resistance	2.5 Ω
L_p	Inductance connected with the inverter	123 μH
V	DC bus voltage	80 V
r_1	ESR of coil on sending side	0.3 Ω
r_2	ESR of coil and capacitance on receiving side	0.17 Ω

Table 2.11 Designed parameters of the WPT system in the design example

Symbol	Parameter	Value
L_1	Resonant inductance on sending side	550.6 μH
L_2	Resonant inductance on receiving side	334.8 μH
k	Coupling coefficient	0.2
R_L	Load resistance	2.5 Ω
L_p	Inductance connected with the inverter	123 μH
V	DC bus voltage	80 V
r_1	ESR of coil on sending side	0.3 Ω
r_2	ESR of coil and capacitance on receiving side	0.17 Ω
C_p	Resonant compensation capacitance	515 nF
r_3	ESR of coil of L_p	0.11 Ω
C_1	Resonant capacitance on sending side	148 nF
C_2	Resonant capacitance on receiving side	189 nF

For the receiving side,

$$C_2 = \frac{1}{\omega^2 L_2} = 189\,\text{nF} \tag{2.123}$$

For the sending side, C_p is given by

$$C_p = \frac{1}{\omega^2 L_p} = 515\,\text{nF} \tag{2.124}$$

Moreover, C_1 can be calculated from Eq. (2.125):

$$C_1 = \frac{1}{\omega^2 (L_1 - L_p)} = 148\,\text{nF} \tag{2.125}$$

Thus, all circuit-component parameters are determined and are listed in Table 2.11.

$$P_{2\text{-}LCC} = I_{2,\text{rms}} V_{L,\text{rms}} = \left(\frac{\omega M V'_{\text{rms}}}{|Z_p Z_S + (\omega M)^2|} \right)^2 R_L = 66.69\,\text{W} \tag{2.126}$$

Based on

$$\eta_{\text{LCC}} = \frac{P_2}{P_1} = \frac{P_2}{\text{real}(i_1) \times V_{\text{rms}}} = 67\% \tag{2.127}$$

2.4 Multi-phase Inverters for High Power

2.4.1 Topology

The proposed multiphase inverter with coupled inductors (namely, InterCell Transformers, ICTs) and blocking capacitors is shown in Fig. 2.24. Figure 2.24b is the equivalent circuit of the inverter shown in Fig. 2.24a. Resistors r_{WIN} and r_{DS} represent the parasitic resistances of ICT windings and the MOSFETs, respectively [3, 13–16].

The self-inductance of every winding is divided into two parts connected in series, namely the magnetizing inductance and the leakage inductance L_{leak}, respectively. The winding self-inductances are L_{1P} through L_{NP} and L_{1S} through L_{NS} at the primary sides and the secondary sides for N ICTs, respectively. The quantity $\dot{Z}_P$ is the total impedance reflected from the receiving side.

The blocking capacitor has two functions. One is to stop the dc currents due to dc-voltage differences between phases. The other is to compensate the voltage drop caused by the leakage inductance. Assuming the blocking capacitor compensates the voltage drop completely, one obtains

$$\omega = \frac{1}{\sqrt{2 L_{\text{leak}} C_{\text{Block}}}} \quad \Rightarrow \quad C_{\text{Block}} = \frac{1}{2\omega^2 L_{\text{leak}}} \tag{2.128}$$

where ω is the switching frequency of the MOSFETs. Additionally, all magnetizing inductances of various ICT windings are supposed to be equal (all windings have the same number of turns). As a result, the circuit of Fig. 2.24b can be transferred to its equivalent one shown in Fig. 2.25, where r_{Block} is the parasitic resistance of the blocking capacitor.

2.4.2 Characteristics and Control

Based on Kirchhoff's law, one obtains the relationship between voltages and currents in each phase of the system presented in Fig. 2.25 as

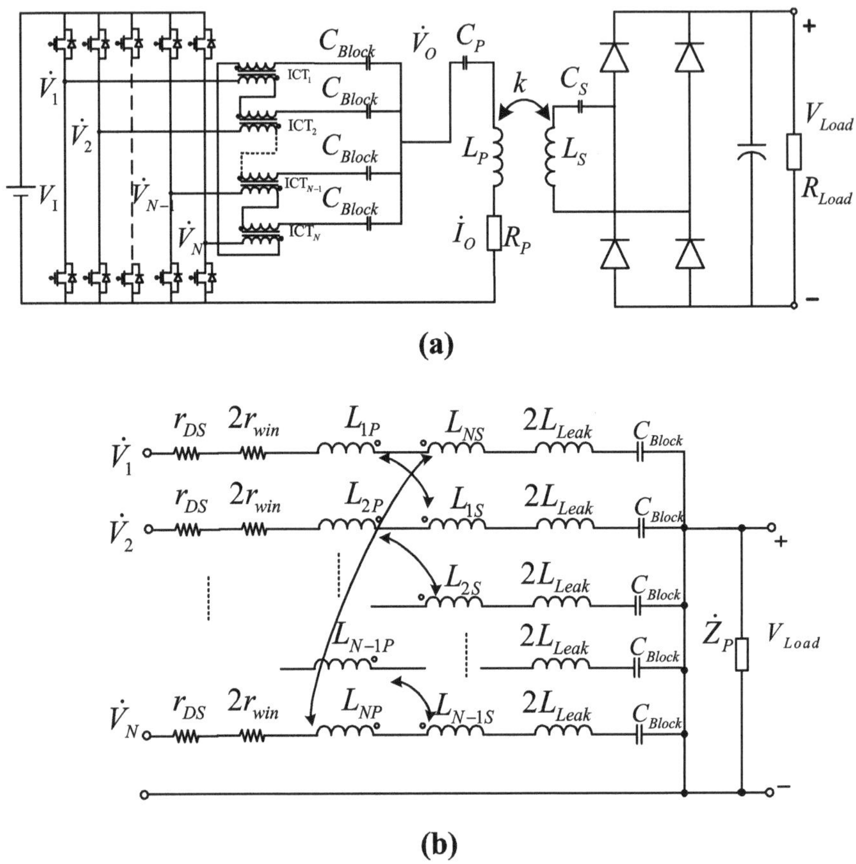

Fig. 2.24 Scheme of multiphase inverter with ICTs and blocking capacitors: (**a**) topology and (**b**) equivalent circuit

$$
\begin{aligned}
\dot{V}_1 - \dot{V}_O &= (2j\omega L_{\text{mag}} + r)\dot{I}_1 - j\omega M_W \dot{I}_N - j\omega M_W \dot{I}_2,\\
\dot{V}_2 - \dot{V}_O &= (2j\omega L_{\text{mag}} + r)\dot{I}_2 - j\omega M_W \dot{I}_1 - j\omega M_W \dot{I}_3,\\
&\vdots\\
\dot{V}_N - \dot{V}_O &= (2j\omega L_{\text{mag}} + r)\dot{I}_N - j\omega M_W \dot{I}_{N-1} - j\omega M_W \dot{I}_1,\\
\dot{V}_O &= \dot{Z}_P \dot{I}_O, \qquad \dot{I}_O = \sum_{i=1}^{N} \dot{I}_i.
\end{aligned}
\tag{2.129}
$$

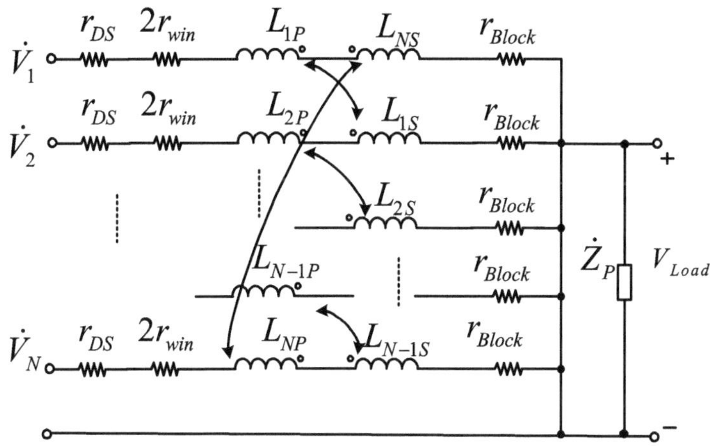

Fig. 2.25 Equivalent circuit of the proposed inverter

where $\dot{I}_1$ through $\dot{I}_N$ are the currents flowing through various phases, and r is the total parasitic resistance of each phase expressed by

$$r = r_{\mathrm{DS}} + 2r_{\mathrm{WIN}} + r_{\mathrm{leak}} + r_{\mathrm{Block}} \tag{2.130}$$

The mutual inductance between two windings of each ICT is given by

$$M_W = k_{\mathrm{mag}}\sqrt{L_{\mathrm{mag}}L_{\mathrm{mag}}} = L_{\mathrm{mag}} \tag{2.131}$$

because the coupling coefficient k_{mag} between magnetizing inductances is assumed equal to unity. Substitution of Eqs. (2.130) and (2.131) into Eq. (2.129) yields the current in each phase of the system:

$$\dot{I}_O = \frac{\sum_{i=1}^{N}\dot{V}_i}{r + N\dot{Z}_P} = \frac{\dot{V}_{\mathrm{avg}}}{r/N + \dot{Z}_P} \tag{2.132}$$

The output voltage is

$$\dot{V}_O = \frac{\dot{Z}_P\sum_{i=1}^{N}\dot{V}_i}{r + N\dot{Z}_P} = \frac{1}{N}\cdot\frac{\sum_{i=1}^{N}\dot{V}_i}{r/(N\dot{Z}_P)+1} = \frac{\dot{V}_{\mathrm{avg}}}{r/(N\dot{Z}_P)+1} \tag{2.133}$$

where $\dot{V}_{\mathrm{avg}}$ is the average of output voltages generated by inverter legs. Considering that the parasitic resistance r is lower than the impedance $\dot{Z}_P$ by several orders of magnitude, Eq. (2.132) can be simplified by ignoring r:

$$\dot{I}_O \approx \frac{\dot{V}_{\mathrm{avg}}}{\dot{Z}_P} \tag{2.134}$$

and the voltage can be expressed as

$$\dot{V}_O \approx \dot{V}_{\text{avg}}. \tag{2.135}$$

One can see from Eq. (2.135) that the output voltage $\dot{V}_O$ is approximately equal to the average of output voltages of multiple phases and is independent of the parameters of the ICT.

When the parasitic resistance r is ignored and the mutual inductance M_W is replaced by L_{mag}, Eq. (2.129) can be rewritten as

$$\dot{\mathbf{V}} = \dot{\mathbf{Z}}\,\dot{\mathbf{I}} \tag{2.136}$$

where

$$\dot{\mathbf{V}} = \begin{bmatrix} \dot{V}_1 - \dot{V}_O \\ \dot{V}_2 - \dot{V}_O \\ \vdots \\ \dot{V}_{N-1} - \dot{V}_O \\ \dot{V}_O \end{bmatrix}, \quad \dot{\mathbf{I}} = \begin{bmatrix} \dot{I}_1 \\ \dot{I}_2 \\ \vdots \\ \dot{I}_{N-1} \\ \dot{I}_N \end{bmatrix}, \quad \dot{\mathbf{Z}} = \begin{bmatrix} j\omega L_{\text{mag}} \begin{bmatrix} 2 & -1 & 0 & \cdots & -1 \\ -1 & 2 & -1 & \cdots & 0 \\ \vdots & \vdots & \vdots & \ddots & \vdots \\ 0 & \cdots & -1 & 2 & -1 \end{bmatrix} \\ \dot{Z}_P \quad \begin{bmatrix} 1 & 1 & 1 \cdots & 1 \end{bmatrix} \end{bmatrix}. \tag{2.137}$$

Hence, currents through various phases can be expressed as

$$\dot{\mathbf{I}} = \dot{\mathbf{Z}}^{-1}\dot{\mathbf{V}}. \tag{2.138}$$

One can define the circulating current as

$$\Delta\dot{I}_i = \dot{I}_i - \frac{\dot{I}_O}{N}, \qquad i = 1, 2, \ldots, N. \tag{2.139}$$

Although the phase currents can be obtained from Eq. (2.138), the resulting expressions are quite complex. Hence, MATLAB codes are developed to find the solution of Eq. (2.139). In fact, all phases shown in Fig. 2.24 are equivalent with regard to circuit topology. Hence, only the circulating current of the first phase is derived. One obtains the circulating current under various parallel phase numbers N from 2 through 6 via MATLAB as

$$\begin{aligned} \Delta\dot{I}_{1,N=2} &= \frac{\dot{V}_1 - \dot{V}_2}{8j\omega L_{\text{mag}}}, \\ \Delta\dot{I}_{1,N=3} &= \frac{2\dot{V}_1 - \dot{V}_2 - \dot{V}_3}{9j\omega L_{\text{mag}}}, \\ \Delta\dot{I}_{1,N=4} &= \frac{5\dot{V}_1 - \dot{V}_2 - 3\dot{V}_3 - \dot{V}_4}{16j\omega L_{\text{mag}}}, \end{aligned} \tag{2.140}$$

$$\Delta \dot{I}_{1,N=5} = \frac{2\dot{V}_1 - \dot{V}_4 - \dot{V}_5}{5j\omega L_{\text{mag}}},$$

$$\Delta \dot{I}_{1,N=6} = \frac{35\dot{V}_1 + 5\dot{V}_2 - 13\dot{V}_3 - 19\dot{V}_4 - 13\dot{V}_5 + 5\dot{V}_6}{72j\omega L_{\text{mag}}}.$$

In extreme cases, all positive terms in the numerator are assumed equal (for example, expressed by $\dot{V}_1$) while all negative terms are kept the same (for example, expressed by $\dot{V}_2$). One then obtains the maximum possible circulating current as

$$\begin{aligned}
\Delta \dot{I}_{1,N=2} &= \frac{1}{8}\frac{\dot{V}_1 - \dot{V}_2}{j\omega L_{\text{mag}}},\\
\Delta \dot{I}_{1,N=3} &= \frac{2}{9}\frac{\dot{V}_1 - \dot{V}_2}{j\omega L_{\text{mag}}},\\
\Delta \dot{I}_{1,N=4} &= \frac{5}{16}\frac{\dot{V}_1 - \dot{V}_2}{j\omega L_{\text{mag}}},\\
\Delta \dot{I}_{1,N=5} &= \frac{2}{5}\frac{\dot{V}_1 - \dot{V}_2}{j\omega L_{\text{mag}}},\\
\Delta \dot{I}_{1,N=6} &= \frac{5}{8}\frac{\dot{V}_1 - \dot{V}_2}{j\omega L_{\text{mag}}}.
\end{aligned} \tag{2.141}$$

Equation (2.141) shows that the maximum possible circulating current increases with the number of parallel phases N.

According to SAE TIR J2954, the 85 kHz band is selected for the light-duty-vehicle wireless charger, which provides an operation-frequency range from 81.38 kHz through 90 kHz. Hence, one selects 88 kHz as the operation frequency of the IPT system. Assuming the rated input voltage of the inverter to be 600 V, one obtains the output-voltage amplitude of phases as 382 V under half-bridge topology with 50% duty. One can analyze the circulating current while assuming the output amplitude of the first phase is equal to the rated one, while the voltage of the second phase varies with respect to its magnitude and angle. The main parameters for circulating-current simulations are listed in Table 2.12. The simulated circulating current based on Eq. (2.141) under various differences of voltage amplitudes and angles for 6 phases is shown in Fig. 2.26. The figure shows that the maximum circulating current is below 0.8 A. As the coefficient for 6 phases is the maximum one of all expressions in Eq. (2.141), 0.8 A is the maximum value for phase numbers from 2 through 6. Keeping in mind that the practical differences of the input dc voltages and phase angle are far below ± 10 V and $\pm 4°$, respectively, the real circulating current will be much smaller than the simulated results. Considering that the coefficient of each expression in Eq. (2.141) is lower than 1, one can derive from Eq. (2.141) that the impedance blocking the circulating currents is higher than ωL_{mag}, which is up to 16 Ω under the parameters listed in Table 2.12. Hence, it

Table 2.12 Parameters for circulating-current simulations

Symbol	Parameter	Value
f	Operating frequency	88 kHz
V_I	Rated input dc voltage of inverters	600 V
V_{I1}	Input voltage of phase 1	V_I
V_{I2}	Input voltage of phase 2	$V_I + \Delta V$
ΔV	Difference of output-voltage amplitude	-10 to 10 V
Φ_1	Output-voltage angle of phase 1	0°
Φ_2	Output-voltage angle of phase 2	$-4°$ to 4°
L_{mag}	Magnetizing inductance of ICT winding	30 μH
N	Number of parallel phases	6

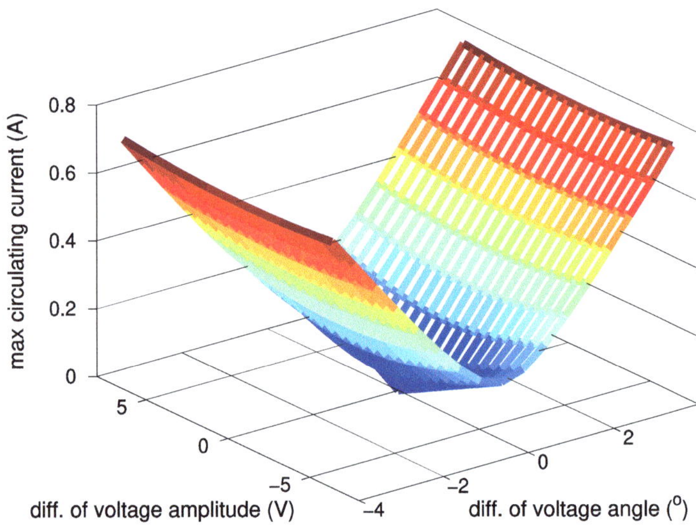

Fig. 2.26 Simulated maximum circulating currents under various differences of input dc voltage and phase angle for the parallel phase number $N = 6$

can be stated that it is the magnetizing inductance of the ICT that provides a high impedance for the differences in phase voltages.

Under a given total output-current amplitude I_O of the inverter with N phases in parallel, the phase current is I_O/N when the current unbalance is ignored. Hence, the total loss due to the MOSFET parasitic resistance r_{DS} is

$$P_{rDS} = N \left(\frac{I_O}{N}\right)^2 \frac{r_{DS}}{2} = \frac{I_O^2 r_{DS}}{2N} \tag{2.142}$$

Equation (2.142) shows that the loss of the parallel topology resulting from r_{DS} is decreased to $1/N$ of the single-phase topology, which provides a possibility of obtaining a higher overall efficiency. However, the introduction of the ICTs leads to extra losses, which may deteriorate the overall efficiency.

Losses at the sending side of the IPT system consist of $P_{r\mathrm{DS}}$, P_{toff}, P_G, $P_{r\mathrm{WIN}}$, $P_{r\mathrm{Block}}$, P_{rL_P}, and P_{rC_P}, which are due to MOSFET drain-source parasitic resistance, MOSFET turn-off, MOSFET drivers, parasitic resistance of one ICT winding, parasitic resistance of the blocking capacitor, parasitic resistance of the resonant inductor, and parasitic resistance of the resonant capacitor, respectively. Assuming that the system operates at the ZVS state, the turn-on loss is ignored. The total loss at the sending side is

$$P_{\mathrm{lossP}} = P_{r\mathrm{DS}} + P_{\mathrm{toff}} + P_G + 2P_{r\mathrm{WIN}} + P_{r\mathrm{Block}} + P_{rL_P} + P_{rC_P} \tag{2.143}$$

where losses related to parasitic resistances ($P_{r\mathrm{WIN}}$, $P_{r\mathrm{Block}}$, P_{rL_P}, and P_{rC_P}) can be calculated via the same method as $P_{r\mathrm{DS}}$ presented in Eq. (2.142). The MOSFET turn-off loss of N phases is expressed as

$$P_{\mathrm{toff}} = \sum_{i=1}^{N} \frac{\omega \left(t_r/3 + t_f/2\right) V_I I_{\mathrm{off},i}}{2\pi} \tag{2.144}$$

where t_r and t_f are the rise time and the fall time of the MOSFET, respectively. The turn-off current $I_{\mathrm{off},i}$ is the current when the MOSFET of the ith phase is turned off and is calculated by

$$I_{\mathrm{off},i} = \frac{I_O}{N} \sin\theta_i \tag{2.145}$$

where θ_i is the angle by which the current lags the voltage of the phase:

$$\theta_i = \arg\left(\frac{\dot{V}_i}{\dot{I}_i}\right), \qquad i = 1, 2, \ldots, N. \tag{2.146}$$

When currents through various phases are balanced, all $I_{\mathrm{off},i}$ are equal, and the same occurs for θ_i. As a result, Eq. (2.144) is simplified to

$$P_{\mathrm{toff}} = \frac{\omega \left(t_r/3 + t_f/2\right) V_I I_O \sin\theta}{2\pi} \tag{2.147}$$

where θ is the angle between the phase current and its voltage. The overall driver loss is

$$P_G = 2NfQ_gV_G \tag{2.148}$$

where Q_g is the gate charge and V_G is the driver voltage of the MOSFET. As a result, Eq. (2.143) is rearranged to

$$P_{\text{lossP}} = \frac{I_O^2(r_{\text{DS}} + 2r_{\text{WIN}} + r_{\text{Block}})}{2N} + \frac{\omega(t_r/3 + t_f/2)V_I I_O \sin\theta}{2\pi} + 2Nf Q_g V_G + \frac{I_O^2 r_{LP}}{2} + \frac{I_O^2 r_{CP}}{2} \tag{2.149}$$

where r_{LP} and r_{CP} are the parasitic resistances of the resonant inductor and capacitor at the IPT sending side, respectively.

The receiving side of the IPT employs a common series-compensation topology and a full-bridge rectifier. The power dissipation at the receiving side is composed of losses P_{VF}, P_{CF}, P_{rL_S}, and P_{rC_S}, which are respectively due to the forward voltage drop of the rectifier diodes, the filter capacitor at the rectifier, the parasitic resistance of the resonant inductor, and the parasitic resistance of the resonant capacitor. The overall loss at the receiving side is

$$P_{\text{lossS}} = P_{VF} + P_{CF} + P_{rL_S} + P_{rC_S} \tag{2.150}$$

where P_{VF} and P_{CF} can be expressed as

$$P_{VF} = \sqrt{2}\, V_F I_S \tag{2.151}$$

$$P_{CF} = \frac{I_S^2 r_{CF}}{2}\left(\frac{\pi^2}{4} - 1\right) \tag{2.152}$$

where V_F, I_S, and r_{CF} are the forward voltage drop of the rectifier diodes, the current amplitude in the resonant loop at the receiving side, and the parasitic resistance of the rectifier filter capacitor, respectively. Hence, Eq. (2.150) can be transferred to

$$P_{\text{lossS}} = \sqrt{2}\, V_F I_S + \frac{I_S^2 r_{CF}}{2}\left(\frac{\pi^2}{4} - 1\right) + \frac{1}{2} I_S^2 \left(r_{L_S} + r_{C_S}\right) \tag{2.153}$$

Let us assume that the single-leg inverter employs the same topology as each phase of the proposed multiphase parallel one (namely, the class-D half-bridge) while the single leg does not use the ICT. As a result, the dissipation of the single-leg topology with the same output current I_O is

$$P_{\text{lossP,single}} = \frac{1}{2} I_O^2 \left(r_{\text{DS}} + r_{LP} + r_{CP}\right) + \frac{\omega\left(t_r/3 + t_f/2\right) V_I I_O \sin\theta}{2\pi} + 2f Q_g V_G \tag{2.154}$$

Table 2.13 Parameters for losses evaluation of the inverters

Symbol	Parameter	Value
f	Operating frequency	88 kHz
r_{DS}	Parasitic resistance of the MOSFET	41 mΩ
r_{WIN}	Parasitic resistance of each ICT winding	40 mΩ
r_{Block}	Parasitic resistance of the blocking capacitor	2 mΩ
Q_g	Charge at the MOSFET gate	300 nC
V_G	Driver voltage of the MOSFET	15 V

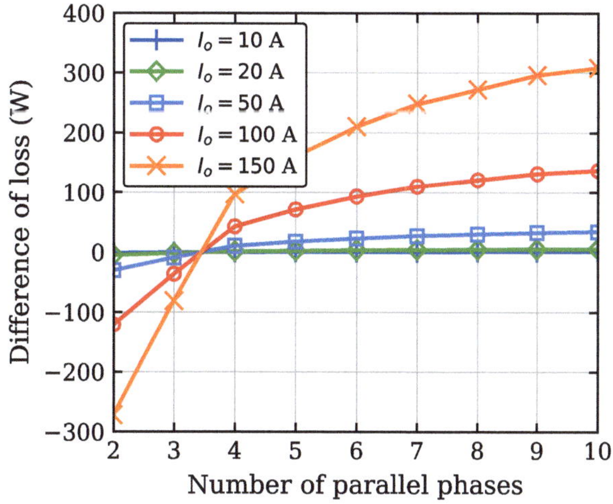

Fig. 2.27 Difference of losses between a single leg and multiphase topologies with various output-current amplitude I

Hence, one obtains the difference of the losses, respectively, for the two topologies with the same output current I_O:

$$\Delta P_{\text{loss}} = P_{\text{lossP,single}} - P_{\text{lossP}} \\ = \frac{I_O^2 r_{\text{DS}}}{2} - \left(\frac{I_O^2 r_{\text{DS}}}{2N} + \frac{I_O^2 r_{\text{WIN}}}{N} + \frac{I_O^2 r_{\text{Block}}}{2N} + 2(N-1) f Q_g V_G \right) \tag{2.155}$$

Under parameters listed in Table 2.13, Fig. 2.27 plots the loss difference presented by Eq. (2.155) between single-phase and multiphase topologies with various numbers of parallel phases. It can be seen that only topologies with more than 3 phases show lower losses than that of a single leg. Furthermore, the higher the phase number is, the bigger the difference of losses becomes. This is because the total resistance (the sum of r_{DS}, r_{WIN}, and r_{Block}) of an N-phase topology is shared by all phases, which leads to an approximately $1/N$ loss of the single leg as shown by Eq. (2.142).

Additionally, topologies with multiple phases do not show loss advantages when the output current is low (e.g., smaller than 20 A). Actually, the terms of Eq. (2.155) can be classified into two parts, namely those due to parasitic resistances and those

Table 2.14 Parameters for the IPT system in Example 2.5

Symbol	Parameter	Value
M	Coupling inductance of two coils	7.33 μH
L_P and L_S	Resonant inductance (sending/receiving)	33.4 μH
C_P and C_S	Resonant capacitance (sending/receiving)	105 nF
r_{C_P} and r_{C_S}	Parasitic resistance of resonant capacitor (sending/receiving)	5 mΩ
r_{L_P} and r_{L_S}	Parasitic resistance of resonant inductor (sending/receiving)	35 mΩ
R_{load}	Load resistance of the rectifier	10 Ω
f	Operating frequency	88 kHz
r_S	On-resistance of MOSFET	5 mΩ
r_{WIN}	ESR of coupling inductor	40 mΩ

due to MOSFET drivers, respectively. At the state with a low current, the overall power dissipation resulting from MOSFET drivers (which stays constant under various currents for a given number of phases) makes up the main part of the whole loss. As a result, the single-phase topology behaves better because of fewer MOSFETs to be driven. However, multiphase topologies perform better with the rise of the output current. It can be seen from Fig. 2.27 that a dissipation saving of 132 W is obtained for a parallel topology with 10 phases when the output-current amplitude goes to 100 A. This is because the losses due to parasitic resistances overwhelm those of the MOSFET driver, while the parallel topology with N legs presents approximately $1/N$ parasitic resistance.

2.4.3 Design Example

Example 2.5 Design a six-phase inverter for a high-power wireless power transfer system employing the topology shown in Fig. 2.24. The WPT system meets the following specifications: the received power P_2 at the load resistance R_{load}, and the circulating current at a maximum phase deviation of 5°. Other parameters are listed in Table 2.14. Evaluate the required input dc voltage, the input power P_1, and the magnetic inductance L_{mag} of the coupled inductor, ignoring the losses in the rectifier and the switching losses in the MOSFETs.

Solution: Via the output current through the resistance,

$$I_{\text{load}} = \frac{1}{\pi}\int_0^{\pi} I_S \sin(\omega t)\, d(\omega t) = \frac{2 I_S}{\pi} \tag{2.156}$$

one obtains the current amplitude at the receiving side:

$$I_S = \frac{\pi I_{\text{load}}}{2} = \frac{\pi\sqrt{P_2/R_{\text{load}}}}{2} = 70.25\ \text{A} \tag{2.157}$$

According to

$$I_P = \frac{Z_2 I_S}{j\omega M} \tag{2.158}$$

the current at the sending side is

$$|I_P| = \left|\frac{Z_2 I_S}{j\omega M}\right| = \left|\frac{\left(j\omega L_S + \frac{1}{j\omega C_S} + \frac{8R_{\text{load}}}{\pi^2}\right) I_S}{j\omega M}\right| = 142.14\,\text{A} \tag{2.159}$$

The input impedance is

$$\begin{aligned} Z &= Z_1 + Z_{\text{ref}} \\ &= j\omega L_P + \frac{1}{j\omega C_P} + \frac{r_{\text{DS}} + r_{\text{WIN}}}{6} + r_{C_P} + r_{L_P} + \frac{(\omega M)^2}{j\omega L_S + \frac{1}{j\omega C_S} + \frac{8R_{\text{load}}}{\pi^2}} \\ &= 2.0256 + j0.9423. \end{aligned} \tag{2.160}$$

The input-impedance angle is

$$\psi = \arctan\left(\frac{\text{Im}(Z)}{\text{Re}(Z)}\right) = 24.94^\circ \tag{2.161}$$

The total power into the inverter is

$$P_1 = \frac{1}{2}\left(I_P^2\left(\frac{r_{\text{DS}} + r_{\text{WIN}}}{6} + r_{C_P} + r_{L_P}\right) + I_S^2(r_{C_S} + r_{L_S})\right) + P_2 = 20.62\,\text{kW} \tag{2.162}$$

When the duty cycle is 0.5 and only the fundamental component is considered, the output power from the inverter is

$$P_1 = \frac{1}{2} \cdot \frac{2V_{\text{DC}}}{\pi} \cdot I_P \cos\psi \tag{2.163}$$

Hence, the required input dc voltage is

$$V_{\text{DC}} = \frac{\pi P_1}{I_P \cos\psi} = 502.87\,\text{V} \tag{2.164}$$

Practically, the real input dc voltage will be a bit higher than the above because of the phase deviation between multi-inverter legs and the losses due to MOSFET switching and the rectifier.

According to the circulating-current expression per Eq. (2.141) for a 6-phase inverter, namely

$$\Delta \dot{I}_{1,N=6} = \frac{5}{8} \frac{\dot{V}_1 - \dot{V}_2}{j\omega L_{\text{mag}}} \tag{2.165}$$

the required magnetizing inductance is

$$L_{\text{mag}} = \frac{5}{8} \left| \frac{\dot{V}_1 - \dot{V}_2}{j\omega \, \Delta \dot{I}_{1,N=6}} \right| = \frac{5}{8} \left| \frac{\frac{2 \times 502.87}{\pi} \left(\sin(\omega t + 5^\circ) - \sin \omega t \right)}{j \, 2\pi \times 88 \times 10^3 \times 2\sqrt{2}} \right| \leq 17.8 \, \mu\text{H} \tag{2.166}$$

Hence, coupling inductors with magnetizing inductance of 17.84 μH can suppress the circulating current to a value below 2 A.

References

1. Q. Deng, J. Liu, D. Czarkowski, M.K. Kazimierczuk, M. Bojarski, H. Zhou, Frequency-dependent resistance of Litz-wire square solenoid coils and quality factor optimization for wireless power transfer. IEEE Trans. Ind. Electron. **63**(5), 2825–2837, 2016. 10.1109/tie.2016.2518126
2. S. Li, Q. Deng, Z. Li, J. Liu, Y. Chang, K. Cui, P. Luo, Identification and H_∞ robust control of wireless power transfer system by Hammerstein model. IEEE Trans. Power Electron. **39**(7), 8883–8893 (2024). 10.1109/tpel.2023.3343085
3. Q. Deng, J. Liu, D. Czarkowski, W. Hu, H. Zhou, An inductive power transfer system supplied by a multiphase parallel inverter. IEEE Trans. Ind. Electron. **64**(9), 7039–7047 (2017). 10.1109/tie.2017.2686351
4. Q. Deng, Q. Huang, Z. Li, H. Deng, S. Li, W. Hu, Ac-resistance evaluation and reduction based on nonuniformity current distribution of large-strand-count Litz wire. IEEE Trans. Power Electron. **40**(7), 10,111–10,122 (2025) 10.1109/tpel.2025.3536849
5. M.K. Kazimierczuk, D. Czarkowski, Class D series-resonant inverters, in *Resonant Power Converters*, 2nd edn., chap. 6 (Wiley-IEEE Press, Hoboken, 2011). ISBN: 978-0-470-90538-8
6. Z. Zhang, H. Pang, A. Georgiadis, C. Cecati, Wireless power transfer—an overview. IEEE Trans. Ind. Electron. **66**(2), 1044–1058 (2019). https://doi.org/10.1109/tie.2018.2835378
7. A. Alabsi, A. Hawbani, X. Wang, A.Y. Al-Dubai, J. Hu, S. Abdel Aziz, S. Kumar, L. Zhao, A.V. Shvetsov, S.H. Alsamhi, Wireless power transfer technologies, applications, and future trends: a review. IEEE Trans. Sustain. Comput. **10**(1), 1–17 (2025). 10.1109/tsusc.2024.3380607
8. V. Ravikiran, R. Keshri, Comparative evaluation of S-S and P-S topologies for wireless charging of electrical vehicles, in *43rd Annual Conference of the IEEE Industrial Electronics Society (IECON)* (2017), pp. 5324–5329. 10.1109/iecon.2017.8216922
9. T. Thabet, J. Woods, Impact of connection type on the efficiency of wireless power transfer systems, in *2017 International Conference on Circuits, System and Simulation (ICCSS)* (2017), pp. 146–150. 10.1109/cirsyssim.2017.8023200
10. Z. Li, Q. Deng, S. Li, H. Deng, Q. Huang, W. Hu, LPV modeling and robust control of dynamic wireless charging for electric vehicles. IEEE Trans. Ind. Electron. **72**(10), 10,629–10,639 (2025). 10.1109/tie.2025.3548993

11. E.S. Lee, S. Lee, T. Phulpin, J.S. Choi, Dipole-type integrated compensation coil design for double-sided LCC topology in EV wireless power transfer systems. IEEE J. Emerg. Selected Topics Power Electron. (2026). 10.1109/jestpe.2025.3650598
12. C.-Y. Chou, M. Tampubolon, J.-Y. Lin, Y.-C. Hsieh, H.-J. Chiu, Study on LCC-C wireless power transfer, in *2017 IEEE Wireless Power Transfer Conference (WPTC)* (2017), pp. 1–4. 10.1109/wpt.2017.7953811
13. Q. Deng, X. Wang, J. Chen, X. Gao, P. Luo, L. Wang, A sclf- balanced multi-channel wireless charging system for EV. IEEE Trans. Veh. Technol. **73**(11), 16,491–16,502 (2024). 10.1109/tvt.2024.3417002
14. M. Bojarski, E. Asa, K. Colak, D. Czarkowski, A 25 kW indus- trial prototype wireless electric vehicle charger, in *2016 IEEE Applied Power Electronics Conference and Exposition (APEC)* (2016), pp. 1756–1761. 10.1109/apec.2016.7468105
15. M. Bojarski, E. Asa, K. Colak, D. Czarkowski, Analysis and con- trol of multiphase inductively coupled resonant converter for wireless electric vehicle charger applications. IEEE Trans. Transp. Electr. **1**(1), 312–320 (2016). 10.1109/tte.2016.2566921
16. Q. Deng, J. Liu, D. Czarkowski, M. Bojarski, E. Asa, F. Leon, Design of a wireless charging system with a phase-controlled inverter for electric vehicles under varying parameters. IET Power Electron. **9**(13), 2461–2470 (2016). 10.1049/iet-pel.2015.0275
17. International Commission on Non-Ionizing Radiation Protection, Guidelines for limiting exposure to time-varying electric and magnetic fields (1 Hz to 100 kHz). Health Phys. **99**(6), 818–836 (2010). 10.1097/hp.0b013e3181f06c86
18. IEEE, *IEEE Standard for Safety Levels with Respect to Human Ex- posure to Electric, Magnetic, and Electromagnetic Fields, 0 Hz to 300 GHz* (Institute of Electrical and Electronics Engineers, 2019). 10.1109/ieeestd.2019.8859679
19. C. Panchal, S. Stegen, J. Lu, Review of static and dynamic wire- less electric vehicle charging system. Eng. Sci. Technol. Int. J. **21**(5), 922–937 (2018). 10.1016/j.jestch.2018.06.015
20. Z. Xue, W. Liu, C. Liu, K.T. Chau, Critical review of wire- less charging technologies for electric vehicles. World Electr. Vehicle J. **16**(2), 65 (2025). 10.3390/wevj16020065

Chapter 3
Class-E Inverters

This chapter examines Class-E resonant inverters for wireless power transfer transmitters. Emphasis is placed on how passive networks shape switch voltage and current waveforms to achieve soft-switching conditions, particularly zero-voltage switching, zero-current switching, and related zero-derivative conditions, to make high-frequency operation possible with low switching loss. The canonical ZVS Class-E inverter is first introduced together with equivalent representations for inductive and capacitive WPT links. Key variants, including ZCS, Class-DE operation, and phase control, are then discussed, followed by a Qi2-oriented design example that connects the analytical relations to a practical system.

3.1 Zero-Voltage Switching

Throughout, we use the following conventions unless otherwise stated: (i) the switching angular frequency is $\omega = 2\pi f$; (ii) $\theta = \omega t$ denotes the normalized time; (iii) the switch device is treated as ideal except where device capacitances and on-resistance are included in the discussion of switching and conduction losses, respectively; and (iv) the WPT link is represented as an equivalent impedance at the inverter output, obtained by reflecting the coupled resonators and rectifier/load to the transmitter side.

3.1.1 Topology

Figure 3.1 shows the canonical Class-E zero-voltage-switching (ZVS) inverter topology as presented in [1]. A large input choke (or RF choke) L_f supplies a

D. Czarkowski et al., *Converter Topologies for Resonant Wireless Power Transfer*, Power Systems, https://doi.org/10.1007/978-3-032-31448-2_3

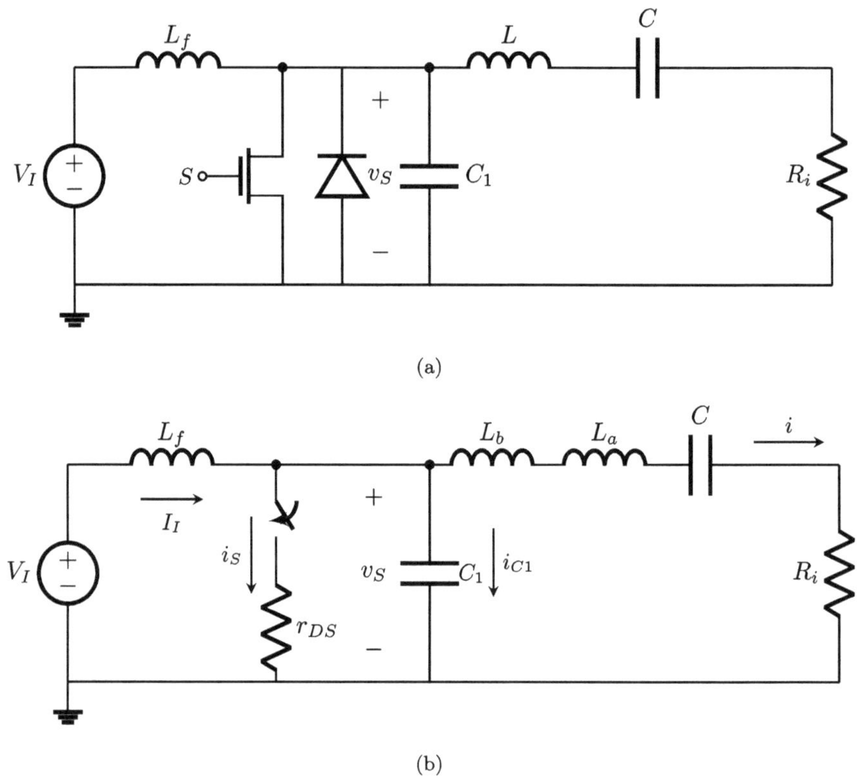

Fig. 3.1 Class E zero-voltage-switching inverter: (**a**) circuit, (**b**) equivalent circuit for operation above resonance with operating (switching) angular frequency $\omega = 1/\sqrt{L_a C}$

nearly constant DC current to the switching node. A shunt capacitance C_1 (often the MOSFET output capacitance plus an external capacitor) shapes the drain-node voltage so that the transistor turns on at (approximately) zero voltage.

A resonant output network (commonly an L–C series network feeding an effective resistance R_{eq}) filters the switching harmonics and produces a near-sinusoidal current at the fundamental frequency. Under the standard high-Q assumption, the current in the series resonant branch can be approximated as a sinusoid

$$i(t) \approx I_m \sin(\omega t + \varphi), \tag{3.1}$$

where φ is the phase of the fundamental current with respect to the fundamental component of the switch voltage.

ZVS/ZDS Switching Conditions
For an ideal Class-E ZVS circuit, the switch drain-to-source voltage $v_S(\theta)$ satisfies, at the turn-on instant,

$$v_S(2\pi) = 0, \tag{3.2}$$

$$\left.\frac{dv_S}{d\theta}\right|_{\theta=2\pi} = 0. \tag{3.3}$$

The first condition is *zero-voltage switching*. The second is often called *zero-derivative switching* (ZDS) (or "zero-voltage-slope switching") and reduces sensitivity to finite device transition times.

Waveforms for Optimum Operation
Figure 3.2 shows the characteristic optimum ZVS/ZDS waveforms: the switch voltage v_S rises during the off interval, peaks significantly above the input voltage, and returns to zero with a zero slope at the next turn-on. The difference of the choke current I_I and the resonant current i is carried alternately by the switch in the ON state as i_S or by the shunt capacitance as i_{C1} during the OFF interval of the transistor.

Key Closed-Form Design Relations at $D = 0.5$
A comprehensive set of closed-form relations for the common design point $D = 0.5$ with a large choke inductance and sufficiently high loaded quality factor is provided in [1]. Defining R_{eq} as the *effective resistive load* seen by the Class-E cell at the fundamental frequency, and V_I as the input DC voltage, the power delivered to R_{eq} in the optimum case can be expressed as

$$P_o \approx 0.5768\,\frac{V_I^2}{R_{\text{eq}}}. \tag{3.4}$$

Equivalently, for a specified P_o and V_I, the optimum resistive load is $R_{\text{eq}} \approx 0.577\,V_I^2/P_o$.

The shunt capacitance required for optimum ZVS/ZDS is commonly written in normalized form as

$$\omega C_1 R_{\text{eq}} \approx 0.1836, \tag{3.5}$$

and the normalized choke-inductance condition for the canonical large-L_f approximation can be expressed as

$$\omega L_f / R_{\text{eq}} = \pi(\pi^2 + 4) \approx 43.6. \tag{3.6}$$

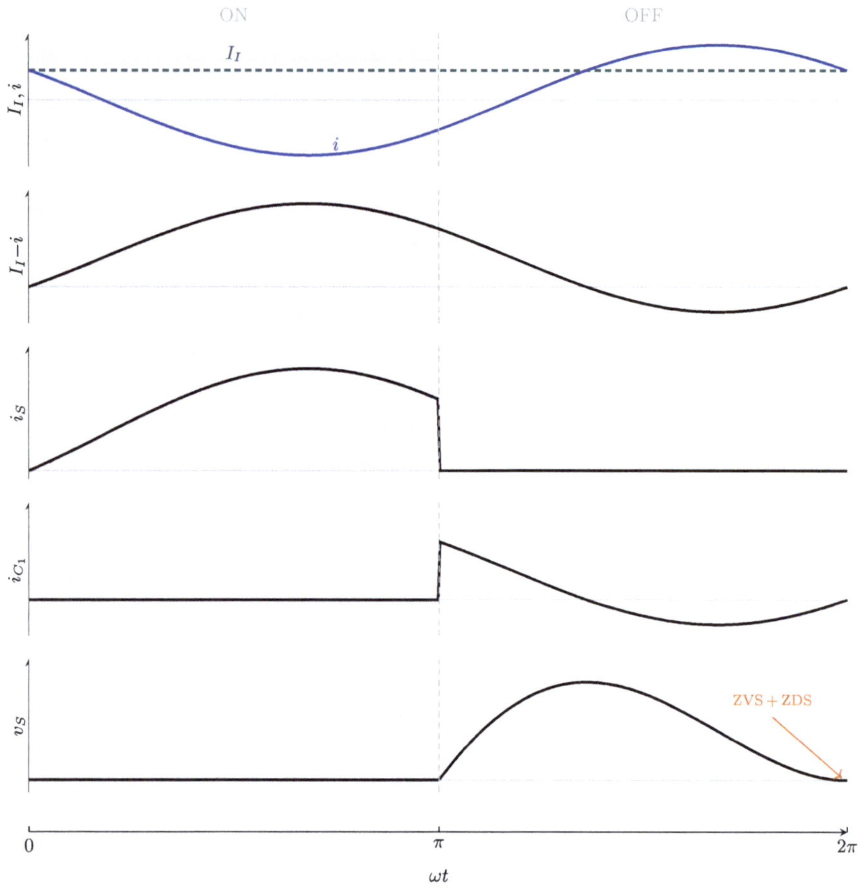

Fig. 3.2 Optimum-operation waveforms for the canonical Class-E ZVS inverter at $D = 0.5$. Gray background marks the switch ON interval. The switch voltage v_S returns to zero with zero slope at turn-on (ZVS+ZDS, arrow)

In this approximation, the choke inductance is chosen large enough that the ac ripple in the input current is small and the input current can be treated as approximately constant. These relations serve as the *nominal* starting point in many practical WPT transmitter designs: once the WPT link and matching circuit are reduced to an equivalent resistance R_{eq} (and possibly a residual reactance), the Class-E inverter can be sized to satisfy (3.2)–(3.3) at the chosen operating frequency.

3.1.2 Characteristics

Voltage and Current Stress
A disadvantage of the Class-E ZVS inverter is high switch voltage stress. For the $D = 0.5$ optimum operation, the peak switch voltage is several times V_I (approximately $3.5V_I$ in the idealized case), which strongly influences MOSFET selection, layout, and snubber strategy. The resonant branch current is approximately sinusoidal and can be related to output power and R_{eq} via $P_o = \frac{1}{2}I_m^2 R_{\text{eq}}$ under the first-harmonic approximation where I_m is the magnitude of the resonant current i.

Soft-Switching Range and "What Varies" in WPT
In a WPT transmitter, the effective load seen by the inverter changes as the coupling changes (misalignment, distance), and as the receiver-side control and rectification change. Maintaining ZVS/ZDS across operating conditions therefore requires (i) keeping the reflected impedance within the range where ZVS holds, and (ii) shaping or matching the impedance so the inverter sees a near-resistive, slightly inductive load at the operating frequency.

To illustrate that, the rest of this section shows two key embeddings of the canonical Class-E ZVS inverter: (i) an *inductive-link (transformer) version* where the coupling coefficient k is explicit in the equivalent impedance, and (ii) a *capacitive WPT (CWPT) version* where coupling is through a capacitance C_M.

3.1.2.1 Transformer (Inductive-Link) Class-E ZVS Inverter with Coupling Coefficient

The transformer/coupled-coil formulation and the persistence of Class-E ZVS under coupling variation are treated in [2, 3].

Equivalent Circuit and Reflected Impedance
Figure 3.3 shows the transformer/coupled-coil representation and the reduction to a single equivalent impedance at the transmitter side [2]. The coupling coefficient is $k \in [0, 1]$, the primary self-inductance is L_p, and for simplification of subsequent equations, it is assumed the secondary inductance is such that $L_s \approx L_p$. The magnetizing and leakage inductances are

$$L_M = kL_p \tag{3.7}$$

$$L_{lp} = (1-k)L_p = \frac{1-k}{k}L_M \tag{3.8}$$

$$L_{ls} = (1-k)L_s \approx (1-k)L_p = \frac{1-k}{k}L_M. \tag{3.9}$$

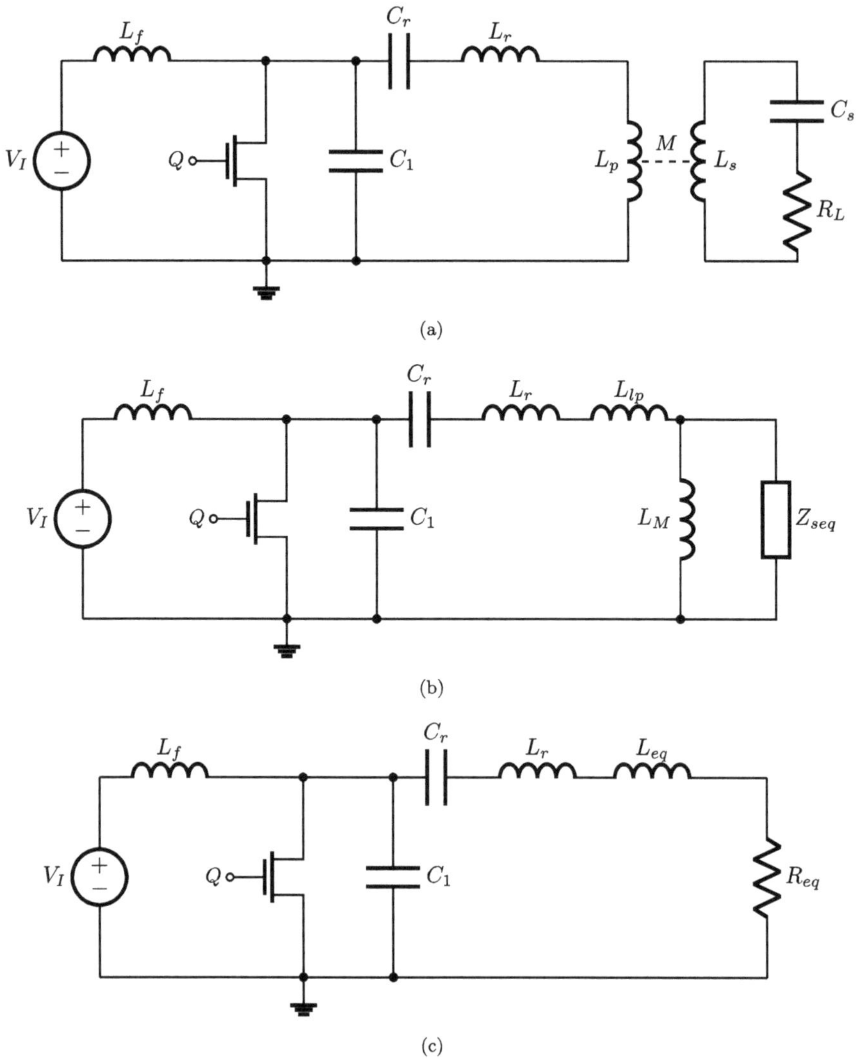

Fig. 3.3 Transformer/coupled-coil Class-E ZVS inverter reduction to an equivalent input impedance

With a secondary compensation capacitor C_s and load R_L, and s standing for jω, the secondary-side impedance is

$$Z_{seq} = sL_{ls} + \frac{1}{sC_s} + R_L. \tag{3.10}$$

The equivalent impedance seen from the primary is

$$Z_{eq} = R_{eq} + \mathrm{j}X_{eq} = sL_{lp} + \left(sL_M \parallel Z_{seq}\right). \tag{3.11}$$

With these assumptions (including $L_s \approx L_p$), evaluating (3.11) at $s = \mathrm{j}\omega$ gives $Z_{eq} = R_{eq} + \mathrm{j}X_{eq}$ with real and imaginary parts expressed in [2] as

$$R_{eq} = \frac{\omega^4 C_s^2 R_L L_p^2 k^2}{\left(1 - \omega^2 C_s L_p\right)^2 + (\omega R_L C_s)^2}, \tag{3.12}$$

$$X_{eq} = \omega L_p \frac{\alpha_0 + \alpha_1}{\beta_0}, \tag{3.13}$$

where

$$\alpha_0 = \omega^4 C_s^2 (1 - k^2) L_p^2, \tag{3.14}$$

$$\alpha_1 = \omega^2 R_L^2 C_s^2 - 2\omega^2 L_p C_s + \omega^2 k^2 L_p C_s, \tag{3.15}$$

$$\beta_0 = \left(1 - \omega^2 C_s L_p\right)^2 + (\omega R_L C_s)^2 . \tag{3.16}$$

The equivalent inductance associated with the reactive part is then

$$L_{eq} = \frac{X_{eq}}{\omega} = L_p \frac{\alpha_0 + \alpha_1}{\beta_0}. \tag{3.17}$$

Finally, the secondary compensation capacitor C_s is chosen to cancel the leakage-inductance reactance,

$$C_s = \frac{1}{\omega^2 L_{ls}} = \frac{1}{4\pi^2 f_s^2 (1 - k) L_p}. \tag{3.18}$$

Practical Range of Coupling Coefficient k

In inductive WPT, k is typically well below unity. Values on the order of $k \sim 0.05$–0.4 are common for loosely coupled systems, depending on geometry and distance; higher values can occur in tightly coupled or well-aligned couplers like in Qi standard phone chargers, lower values are typical in EV charging. Equations (3.12)–(3.18) make the role of k explicit:

- $R_{eq} \propto k^2$ (all else equal), so delivered power and switch stresses change sharply with coupling.
- The reactive part depends on both $(1 - k^2)$ and the compensation choice; imperfect cancellation of leakage/reactance can move the Class-E cell away from its ZVS/ZDS point.

A complementary coupling-aware design viewpoint is developed in [3]: the inverter is sized for a selected maximum coupling so that both ZVS and zero-derivative switching (ZDS) are satisfied. The analysis further shows that ZVS is maintained as coupling decreases, provided the magnetizing and leakage inductances are absorbed into the effective network. For consistency, the equation set and notation used here follow [2].

Matching Network for Imposing the Canonical Class-E Conditions

Once the coupled-coil and receiver are reduced to an equivalent impedance, the remaining design task is to: (i) cancel the reactive part so the Class-E resonant branch sees an effective resistance, and (ii) transform that effective resistance to the optimum R_{eq} required by (3.4)–(3.6).

Reference [2] carries this out by selecting the canonical Class-E component relations (written here in the same form as for $D = 0.5$ optimum operation in [1]); C_1 denotes the total shunt capacitance across the switch, including the MOSFET output capacitance, any external parallel capacitor, and, where relevant, the choke parasitic/self-capacitance and stray capacitances:

$$R_{\mathrm{eq}} = \frac{8}{\pi^2+4}\frac{V_I^2}{P_o}, \tag{3.19}$$

$$L = L_r + L_{eq} = \frac{Q_L R_{\mathrm{eq}}}{\omega}, \tag{3.20}$$

$$C_1 = \frac{8}{\pi(\pi^2+4)\omega R_{\mathrm{eq}}}, \tag{3.21}$$

$$C_r = \frac{1}{\omega R_{\mathrm{eq}}\left(Q_L - \frac{\pi(\pi^2-4)}{16}\right)}, \tag{3.22}$$

$$L_f = \frac{2\left(\pi^2/4+1\right)R_{\mathrm{eq}}}{f_s}. \tag{3.23}$$

A π-type matching network can then be used to realize the needed impedance transformation. Figure 3.4a shows the abstract $\pi 2a$ matching network, while Fig. 3.4b shows its correspondence to the Class-E WPT circuit. At the nominal tuned operating point, the abstract network variables are identified as

$$L_1 = L_r + L_{lp}, \qquad L_2 = L_M, \qquad R_i \approx Z_{seq}(j\omega) \approx R_L,$$

so the matching reactances satisfy

$$X_{L1} = \omega L_1 = \left(Q_L - \sqrt{\frac{R_i}{R_{\mathrm{eq}}} - 1}\right) R_{\mathrm{eq}}, \tag{3.24}$$

$$X_{L2} = \omega L_2 = \frac{R_i}{\sqrt{\frac{R_i}{R_{\mathrm{eq}}} - 1}}, \tag{3.25}$$

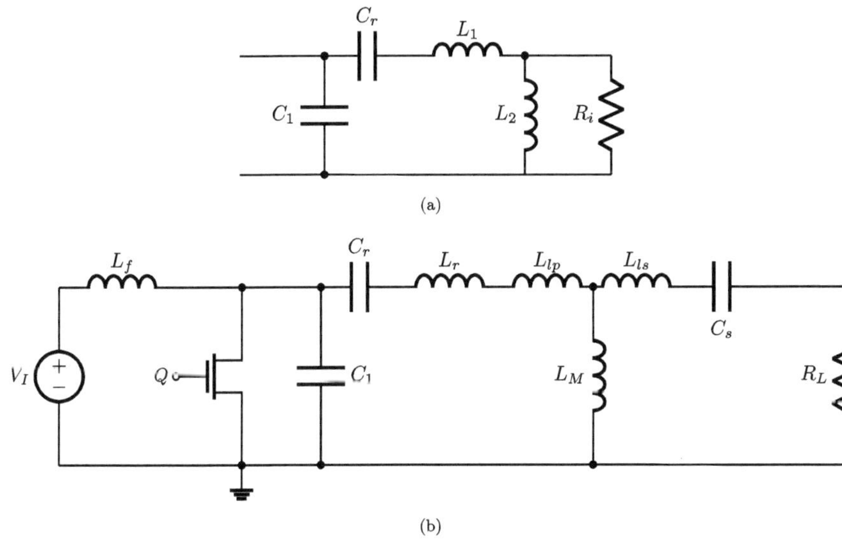

Fig. 3.4 Matching resonant circuit used to transform the coupled-link/load to the desired Class-E equivalent resistance

$$R_{eq} = \frac{R_i}{1 + \left(\frac{R_i}{X_{L2}}\right)^2}. \tag{3.26}$$

Thus, Eqs. (3.24)–(3.26) describe the abstract $\pi 2a$ matching network of Fig. 3.4a, whereas the identifications $L_1 = L_r + L_{lp}$ and $L_2 = L_M$ connect that network to the physical Class-E WPT circuit in Fig. 3.4b. Together, (3.7)–(3.26) provide a practical design flow: choose (f_s, P_o) and a *nominal* coupling k (and receiver R_L), compute the equivalent impedance parameters, pick Q_L, and synthesize the matching so that the Class-E cell operates near its canonical ZVS/ZDS conditions at that nominal point.

3.1.2.2 Capacitive WPT Class-E ZVS Inverter

In capacitive WPT (CWPT), the coupling element is a capacitance between plates rather than a mutual inductance between coils. The plate-level realization and a detailed derivation of the CWPT equivalent circuit are given in [4]. Figure 3.5 shows a reduced equivalent circuit of the full plate-level CWPT system.

In the reduced model, the CWPT section is replaced by

$$Z_{eq} = R_{eq} + jX_{eq}, \tag{3.27}$$

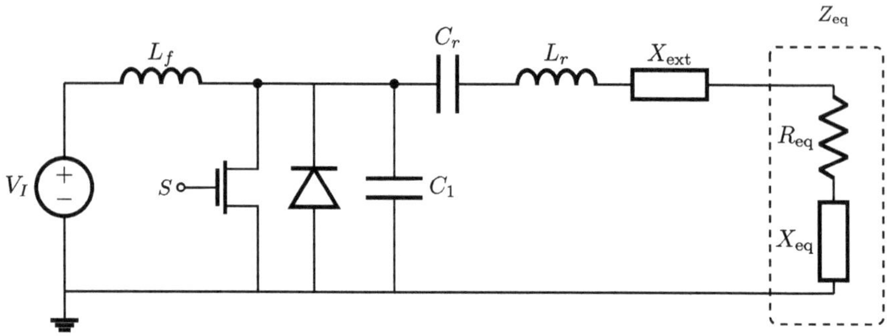

Fig. 3.5 Reduced Class-E CWPT inverter model. The capacitive coupling network is represented by the equivalent impedance $Z_{\text{eq}} = R_{\text{eq}} + \mathrm{j}X_{\text{eq}}$, while X_{ext} denotes the added series reactance used to cancel the residual reactive part at the nominal operating point

where both R_{eq} and X_{eq} depend on coupling and load. A nominal coupling point $C_{M,\text{opt}}$ is then selected, and the residual reactive component is canceled by an added series reactance,

$$X_{\text{ext}} = -X_{\text{eq}}. \tag{3.28}$$

At that nominal point, the Class-E cell sees a purely resistive equivalent load, and the remaining resonant elements can be tuned for ZVS/ZDS operation using the closed-form procedure in [4].

Interpretation for WPT Robustness

The reduced CWPT representation makes the central design message clear: the geometric and material properties of the coupling structure need not be carried through the Class-E analysis. Instead, the link is represented by an equivalent impedance Z_{eq}. As plate spacing or alignment varies, this impedance shifts, changing the effective load seen by the inverter and thus its soft-switching range.

Recent Exact-Analysis and Robustness Refinements

A recent WPT-oriented study [5] revisits the Class-E inverter without imposing the usual high-Q sinusoidal approximation on the resonant variables. The resulting exact time-domain solution expresses the key waveforms as combinations of sinusoidal and exponential terms, which is useful when waveform distortion is significant or when Q_L is too low for the canonical approximation to hold. Experimental verification shows that this more exact viewpoint can complement the standard closed-form Class-E design flow in practical WPT analysis.

A complementary practical-design perspective is provided in [6], which develops a synthesis methodology for Class-E and related single-switch inverters that preserves ZVS over a wide range of effective resistive loads. Although the paper is not written specifically for WPT, its core message is directly relevant here: coupling variation, misalignment, and receiver-side operating changes appear at the transmitter as load modulation. Hence, robustness of the Class-E cell to load-resistance variation is both desirable and, as that paper demonstrates, achievable.

3.2 Variants of Class E

The canonical Class-E ZVS inverter is only one member of a broader family of "waveform-engineered" resonant inverters. This section addresses three prominent variants used in WPT: (i) zero-current-switching (ZCS) circuits, (ii) Class-DE circuits with operation that blends Class-D structure with Class-E-like capacitive shaping, and a third variant—phase control at fixed frequency—that is discussed in Sect. 3.2.3.

3.2.1 Zero-Current Switching

Class-E zero-current-switching (ZCS) inverters replace the ZVS turn-on condition for a *zero-current* (and often zero-current-derivative) condition at a switching transition. In the canonical ZCS circuit, the switch is arranged to *turn off* at zero current, which eliminates turn-off loss and reduces voltage spikes at turn-off [1]. In contrast to ZVS, the basic ZCS topology does not inherently absorb the transistor output capacitance, so the switch may turn on at a nonzero voltage, and turn-on loss can become dominant at very high frequencies unless additional measures are taken.

Canonical Topology and Equivalent Circuit
Figure 3.6 shows the canonical Class-E ZCS inverter and its equivalent circuit [1]. The series network is represented by an effective resistance R_i at the fundamental component and a resonant path in which the net series capacitance C is split into C_a and C_b. This partition emphasizes that C_a resonates with L at the operating frequency; the additional capacitance C_b reflects that the full L–C branch operates below its natural series-resonant frequency.

Resonant Relationships and Loaded Quality Factor
With L resonating with C_a at frequency $f = \omega/(2\pi)$, the angular operating frequency is

$$\omega = \frac{1}{\sqrt{LC_a}}. \tag{3.29}$$

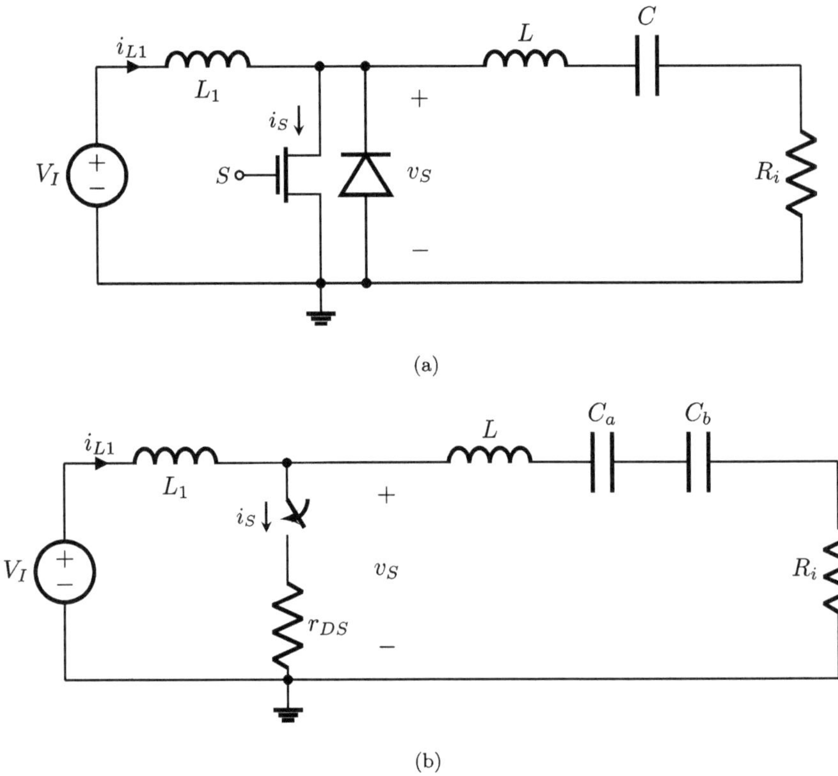

Fig. 3.6 Canonical Class-E ZCS inverter topology

Let X_{C_r} denote the magnitude of the net series capacitive reactance seen in the L–C–R_i path at ω. The loaded quality factor is defined as

$$Q_L = \frac{X_{C_r}}{R_i} = \frac{C_a + C_b}{\omega R_i C_a C_b}. \tag{3.30}$$

The parameter Q_L influences stresses, harmonic content, and detuning sensitivity—all of which become system-level issues in WPT because coupling and load variation change the reflected impedance.

Switching Intervals and Soft-Switching Conditions

For the baseline analysis, let us assume a 50% duty ratio: the switch is OFF for $0 < \theta \le \pi$ and ON for $\pi < \theta \le 2\pi$, consistent with the waveform convention used in Fig. 3.7. Let $i_S(t)$ denote the switch current. The ZCS boundary condition at turn-off is

$$i_S(2\pi) = 0. \tag{3.31}$$

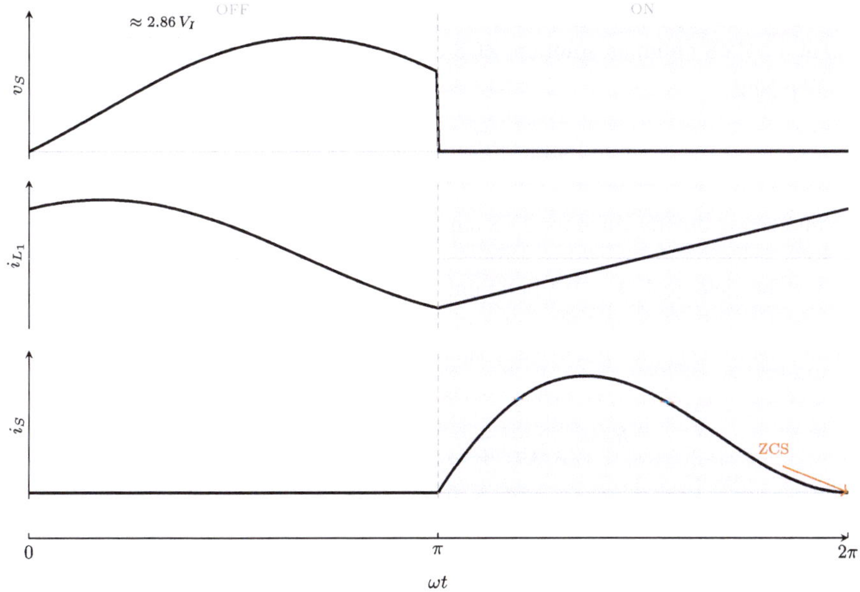

Fig. 3.7 Optimum-operation waveforms for the Class-E ZCS inverter at $D = 0.5$. Gray background marks the switch ON interval ($\pi \le \omega t \le 2\pi$). The switch voltage peak $\approx 2.86\,V_I$ is notably lower than the ZVS counterpart ($\approx 3.56\,V_I$); the switch current vanishes at turn-off (ZCS, arrow)

Optimum ZCS operation is defined by additionally enforcing a zero current derivative at the switching instant,

$$\left.\frac{\mathrm{d}i_S}{\mathrm{d}\theta}\right|_{\theta=2\pi} = 0, \tag{3.32}$$

which keeps $i_S(\theta) \ge 0$ and prevents conduction of the antiparallel diode over the cycle.

Optimal Waveforms

Figure 3.7 shows the characteristic optimum waveforms of the Class-E ZCS inverter at $D = 0.5$ [1]. During the OFF interval, $i_S = 0$ and the inductor/tank current approximately equals the output current; during the ON interval, $v_S = 0$ and the inductor current ramps. The waveforms illustrate the key tradeoff: turn-off occurs at zero current, but the topology does not inherently absorb the switch output capacitance, so turn-on need not be at zero voltage.

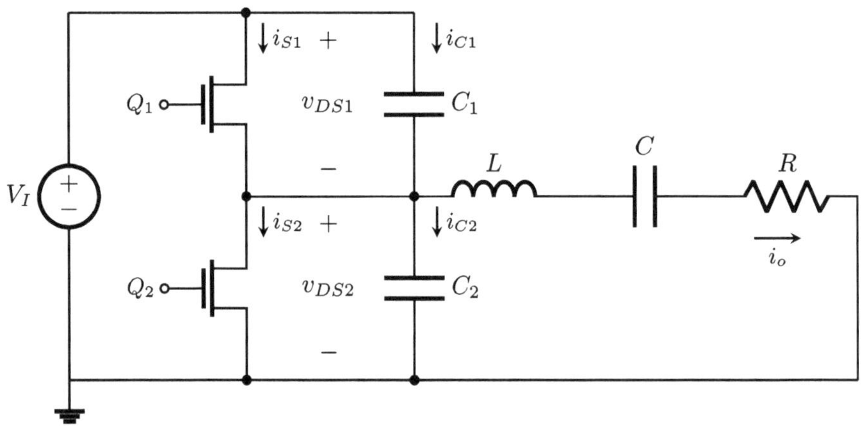

Fig. 3.8 Half-bridge Class-DE inverter. Voltage probes v_{DS1}, v_{DS2} and current arrows i_{S1}, i_{S2}, i_{C1}, i_{C2}, i_o correspond directly to the waveform panels in Fig. 3.10

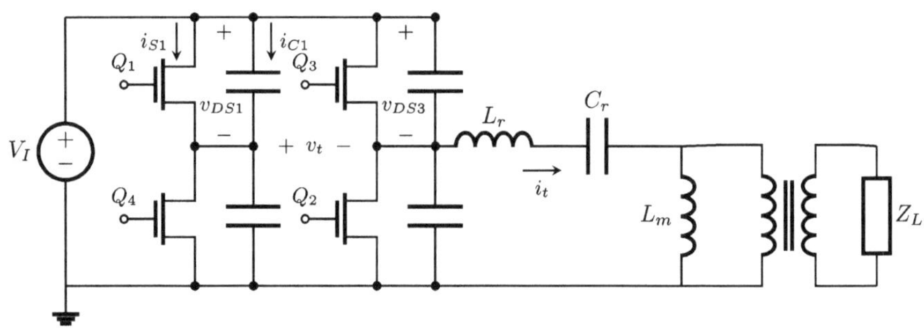

Fig. 3.9 Full-bridge Class-DE inverter with connection to WPT load

Classic Half-Bridge and Full-Bridge Extension

Figure 3.8 shows the half-bridge Class-DE topology [1], while Fig. 3.9 shows a full-bridge implementation and associated WPT resonant network and load arrangements [11]. The full-bridge form is often preferred in higher-power WPT transmitters because it: (i) doubles the available fundamental excitation for a given DC bus, and (ii) reduces even-harmonic content and certain common-mode stresses when properly balanced.

Waveforms and Soft-Switching Mechanism

In Class-DE operation, the MOSFETs are driven with a dead time so that, during commutation, the resonant tank current charges and discharges the MOSFET output capacitances (or explicit shunt capacitors), bringing the device voltage to zero before turn-on. This is reflected in the typical waveforms shown in Fig. 3.10 (half-bridge) and Fig. 3.11 (full-bridge): the device-voltage transitions occur within the dead-

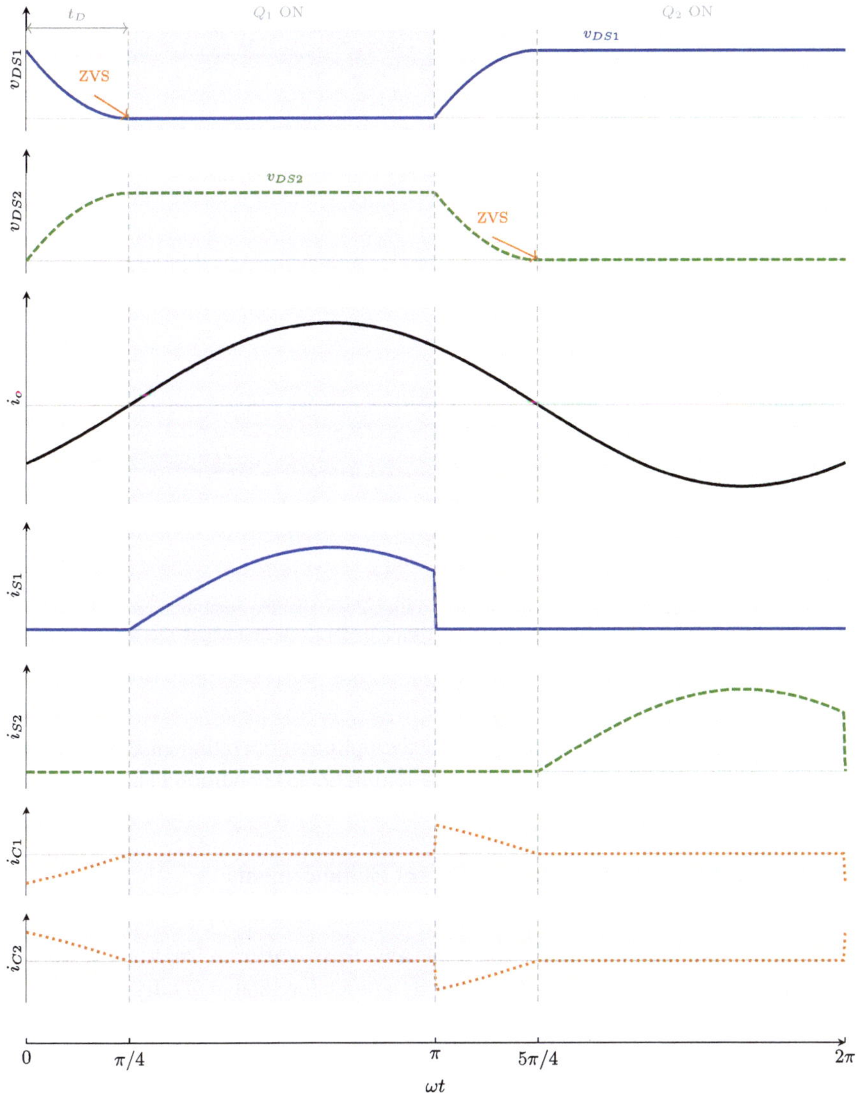

Fig. 3.10 Half-bridge Class-DE waveforms at $D = 0.375$. Color coding matches Fig. 3.8: orange = dead-time t_D (capacitors C_1, C_2 drive ZVS transitions); blue = Q_1 ON; green = Q_2 ON. Orange arrows mark the ZVS turn-on instants where v_{DS} reaches zero

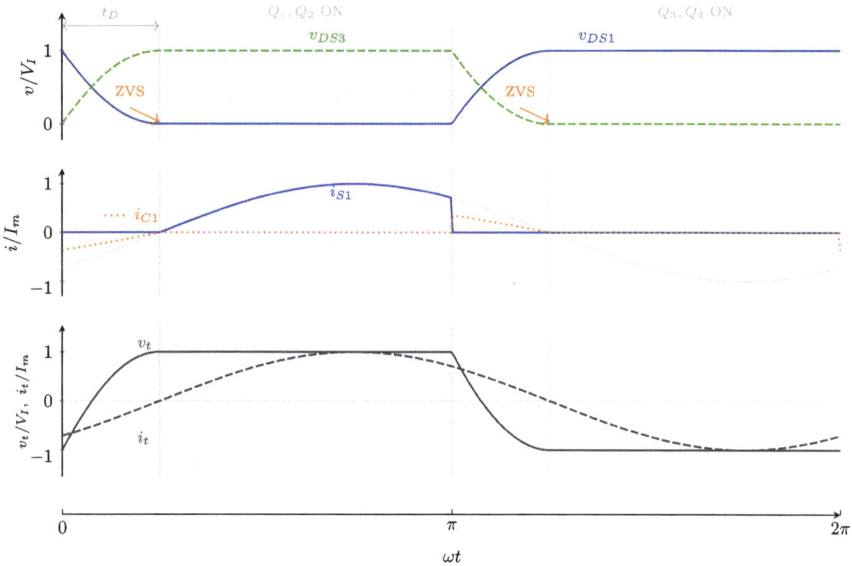

Fig. 3.11 Full-bridge Class-DE waveforms illustrating ZVS-enabled commutation. Color coding: blue = $S_{1,2}$ ON; green = $S_{3,4}$ ON; orange = dead-time intervals during which the shunt capacitors C_{sw} provide the ZVS voltage transition

time intervals, enabling ZVS turn-on. For the optimum operating case, the turn-on instants also satisfy the zero-derivative-switching (ZDS) condition [1].

WPT-oriented Seminal Works and Recent Enhancements

The following WPT-oriented Class-DE results are relevant:

- **Steady-state analysis and design at any duty ratio:** A rigorous Class-DE steady-state model and closed-form design procedure that explicitly covers duty-ratio variation and MOSFET parasitic capacitances is developed in [17]. The paper shows that the highest maximum operating frequency occurs at $D = 0.25$ and recommends the practical range $0.25 \leq D < 0.5$.
- **Control by phase shift:** Fixed-frequency phase control for a Class-DE/E resonant DC–DC converter, with output proportional to phase shift, is developed in [18].
- **Wide-impedance matching:** An impedance-matching method that maintains high efficiency over a wide impedance range is proposed in [19].

These enhancements address a central WPT challenge: the reflected impedance is not constant. The choice of half- or full-bridge structure, explicit commutation capacitors, and matching/control strategies give Class-DE considerable design flexibility across power levels. For WPT-specific co-design of a Class-E inverter and Class-DE rectifier, see also the related ISCAS study in [20].

3.2.3 Phase Control

Fixed-frequency control is often preferred in WPT for EMI management, filter simplification, and predictable resonant behavior. Phase control (also known as outphasing modulation in RF contexts) achieves output-power regulation at a fixed switching frequency by combining two synchronized inverter "cells" whose fundamental outputs are shifted in phase relative to each other.

Current-Source vs. Voltage-Source Phase-Controlled Concepts
Two fundamental phase-control archetypes can be distinguished [1]:

- **Current-source phase control:** two sinusoidal current sources deliver a vector-summed output current. If the amplitudes are equal ($I_{m1} = I_{m2} = I_m$), the output-current amplitude becomes

$$I_{om} = 2I_m \cos\left(\frac{\phi}{2}\right), \tag{3.37}$$

  so output power decreases as ϕ increases from 0° to 180°.
- **Voltage-source phase control:** two sinusoidal voltage sources are combined differentially to form the output voltage. If the amplitudes are equal ($V_{m1} = V_{m2} = V_m$), the output-voltage amplitude becomes

$$V_{om} = 2V_m \sin\left(\frac{\phi}{2}\right), \tag{3.38}$$

  so output power increases as ϕ increases from 0° to 180°.

Class-E Phase-Controlled Inverter Example
Figure 3.12 shows an example of a Class-E phase-controlled voltage-source inverter constructed from two Class-E ZVS cells [1] where R_L represents the WPT load. In this architecture, the matching networks make each Class-E cell behave like a sinusoidal voltage source at the fundamental, and the phase shift ϕ becomes the control variable.

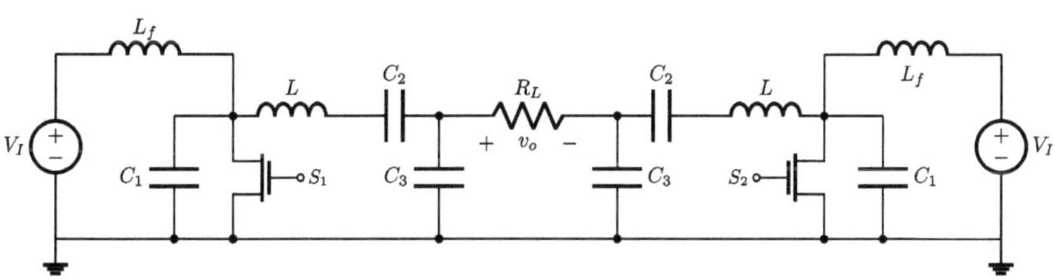

Fig. 3.12 Example of a Class-E phase-controlled voltage-source inverter

Reference [1] gives a useful ZVS feasibility condition in terms of the effective load resistance. If $R_{L,\text{opt}}$ denotes the nominal optimum load of a single Class-E cell, phase control changes the effective load presented to each cell. For the voltage-source case, the ZVS condition becomes

$$R_L > 2R_{L,\text{opt}} \sin^2\left(\frac{\phi}{2}\right), \tag{3.39}$$

illustrating that phase shift can move the operating point into or out of the ZVS range.

Asymmetric Phase-Controlled Class-E Design

The design method in [21] addresses a key limitation of symmetric phase-controlled Class-E inverters: a non-ZVS region can appear as ϕ deviates from the nominal design state. An asymmetric configuration and design method are proposed so that both Class-E cells satisfy ZVS continuously over the control range. The asymmetry is introduced by making the component values depend on a nonzero nominal phase shift. Figure 3.13 shows the corresponding asymmetric circuit configuration with two Class-E cells feeding a common load through separate resonant branches, while Fig. 3.14 shows representative switching and output waveforms.

Remark on ZDS in WPT Operation

In canonical Class-E design, ZDS ($dv_S/d\theta|_{\text{on}} = 0$, equivalently $dv_{DS}/d(\omega t)|_{\text{on}} = 0$) is imposed together with ZVS at the nominal operating point. In a WPT inverter, variations in coupling, load, or control phase modify the reflected impedance and hence the switch-current waveform, so ZDS is generally not preserved away from that nominal point, even though ZVS may still be maintained over part of the operating range.

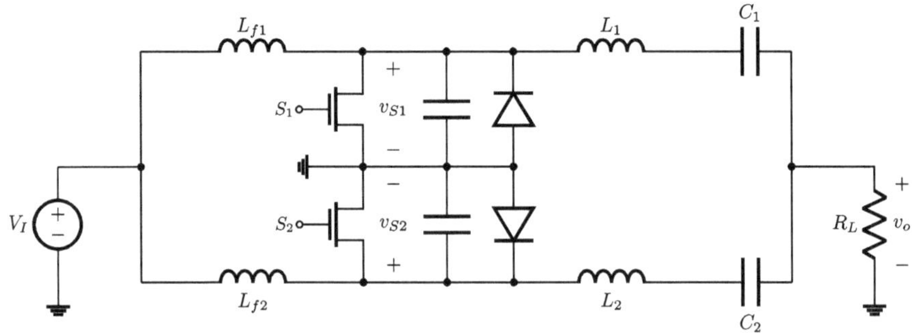

Fig. 3.13 Asymmetric phase-controlled Class-E inverter configuration

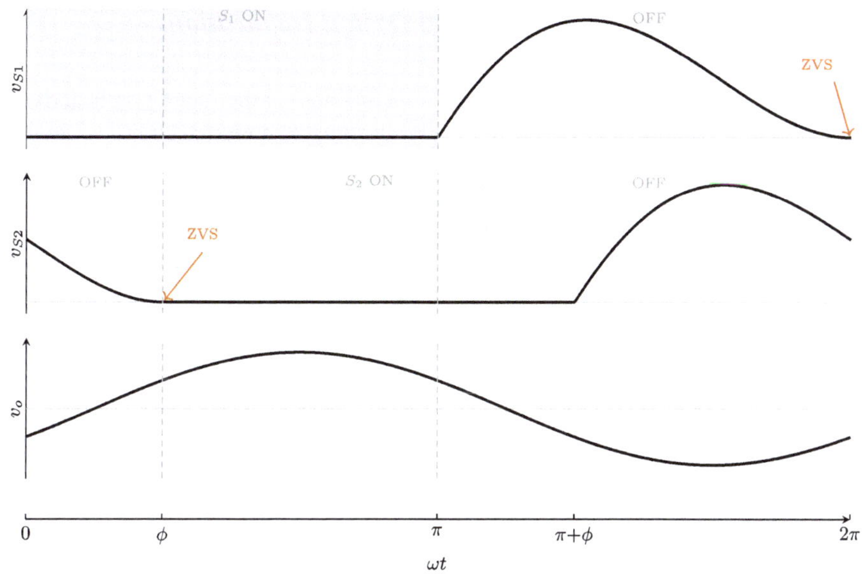

Fig. 3.14 Phase-controlled Class-E inverter waveforms

Connections to WPT Regulation

Phase control is one of several fixed-frequency regulation approaches for WPT systems. Three recent papers illustrate complementary fixed-frequency strategies and broader context:

- An adaptive impedance-matching network implemented as a selectable capacitor array maintains near-optimum Class-E operation under coupling variation without changing frequency [22].
- A bidirectional Class-E inverter/rectifier WPT model uses phase shift between transmitter and receiver switching signals to control power-flow direction and magnitude [23].
- A wireless DC motor drive example combines an interleaved wide-load Class-E transmitter with a receiver-side full-bridge phase-shift stage, illustrating that "phase shift" can appear at different points in the overall WPT power chain [24].

These works reinforce a recurring theme: *fixed-frequency* operation motivates control through phase, impedance transformation, or multi-cell vector combination, rather than frequency modulation. A recent industrial-scale example [25] further shows that Class-E-derived fixed-frequency power-combining concepts can be extended to the multi-kW, multi-MHz range through a parallel differential Class-E architecture and off-nominal design of the coupled system.

3.2.4 Comparative Summary and Design Implications

Table 3.1 summarizes the main practical tradeoffs among the four Class-E-derived approaches discussed in this chapter. In WPT design, the preferred choice is usually dictated by semiconductor characteristics (especially the available C_{oss} and reverse-recovery behavior), the required power level, and whether regulation is achieved primarily through impedance shaping, frequency variation, or fixed-frequency phase control.

A recurring message throughout this monograph is that, in WPT, inverter design and link design should be conducted concurrently: the inverter is designed for an *equivalent impedance* that depends on coupling, compensation, and rectification. Accordingly, modern research increasingly combines impedance-conditioning networks, load-independent synthesis, and fixed-frequency control so that the desired switching boundary conditions can be preserved under realistic operating variations.

Recent exact-analysis work on Class-E WPT inverters [5] and recent multi-kW demonstrations of parallel differential Class-E WPT systems [25] suggest that future comparisons should distinguish more clearly between canonical single-cell design theory and scalable multi-cell implementations under off-nominal operation. In that sense, the four families summarized in Table 3.1 should be viewed not as isolated

Table 3.1 High-level comparison of Class-E-derived inverter families for WPT

Family	Soft-switching target	Typical strengths in WPT	Common caveats
Class-E ZVS	ZVS (+ZDS) at turn-on	Excellent at MHz with significant C_{oss}; canonical closed-form sizing; high efficiency when matched	High peak switch voltage; soft-switching margin depends on reflected impedance phase
Class-E ZCS	ZCS (+zero slope) at turn-off	Reduced turn-off loss and voltage stress; compatible with some parallel-resonant links	Turn-on can be hard (nonzero voltage); switch output capacitance is not naturally absorbed, so additional measures may be needed at high frequency
Class-DE	ZVS (+ZDS at the optimum point) during dead time	Bridge scalability and symmetry; lower device voltage stress than single-ended Class-E; explicit shunt capacitors make commutation design transparent	Requires sufficiently inductive input impedance for proper commutation; dead-time design is sensitive to impedance drift
Phase control	Fixed-f regulation via vector combination	Fixed-frequency regulation while preserving resonant tuning; can maintain soft switching with proper design	ZVS region can shrink with phase shift; may require asymmetric design or adaptive matching

alternatives, but as design templates whose suitability depends on how the overall WPT system is tuned, controlled, and scaled.

3.3 Design Example: 20 W Fixed-Frequency Class-E Transmitter at 360 kHz

This section uses the impedance-reduction approach and the Class-E ZVS design relations from Sect. 3.1 to work through a representative consumer-WPT transmitter design in the 360 kHz, 20 W class, motivated by Qi2-type operating conditions [26]. The goal is a step-by-step design that goes from closed-form relations to component values, suitable for simulation and hardware verification. Class-E ZVS is used here because, at 360 kHz and 15 V, the MOSFET output capacitance can be absorbed naturally into the required shunt capacitance, and the canonical closed-form relations provide a direct single-switch design flow. At the nominal coupling point, the coupled-coil reflected impedance is approximately resistive, so the standard Class-E assumptions are a reasonable first approximation. For this low-voltage, moderate-power example, ZCS does not offer a clear advantage, while a Class-DE realization would require at least a two-switch structure whose additional complexity is not needed here. A Qi-standard Class-E design at 100 kHz and 15 W, including EMI filter and hardware verification, is presented in [27].

3.3.1 Specifications

- Output power: $P_o = 20\,\text{W}$.
- DC input voltage: $V_I = 15\,\text{V}$.
- Switching frequency: $f_s = 360\,\text{kHz}$ ($\omega = 2\pi f_s \approx 2.26 \times 10^6\,\text{rad/s}$).
- Coupling design point: $k_{\max} = 0.7$ (a representative value for a closely aligned consumer-WPT coil pair, used as the nominal design point).
- Coils (example): $L_1 = L_2 = 24\,\mu\text{H}$.

3.3.2 Main Assumptions and Modeling Notes

Main Assumptions (First-Cut Sizing)

1. **Fundamental-harmonic approximation (FHA):** the resonant tank and reflected load are analyzed at the fundamental frequency ω only.
2. **Tuned receiver at** ω: the secondary port presents $Z_2(j\omega) \approx R_L$ at the design frequency.

3. **Ideal elements in the nominal sizing step:** coil ESR, switch $r_{ds,on}$, and nonlinear $C_{oss}(v)$ are neglected in the first-cut design; their effect is assessed using the parasitic checklist in Sect. 3.3.5.
4. **Constant coupling coefficient:** $M = k\sqrt{L_1 L_2}$ with fixed k (no core saturation and no proximity-effect detuning).

3.3.3 Component Calculation

Step 1: Optimum Equivalent Resistance
From (3.4), the optimum fundamental resistive load seen by the Class-E cell is

$$R_{\text{eq}} \approx 0.5768\,\frac{V_I^2}{P_o} = 0.5768\,\frac{15^2}{20} \approx 6.49\,\Omega. \tag{3.40}$$

Step 2: Coupling-Dependent Load Target
Let the mutual inductance be

$$M = k\sqrt{L_p L_s}.$$

With the secondary tuned to resonate at ω (i.e., $C_s = 1/(\omega^2 L_{ls})$ so that $Z_2(j\omega) \approx R_L$), the reflected resistance is

$$R_{\text{eq}} \approx \frac{(\omega M)^2}{R_L} \quad \Longrightarrow \quad R_L \approx \frac{(\omega M)^2}{R_{\text{eq}}}. \tag{3.41}$$

Validity note: Under the FHA used here, (3.41) introduces no additional error when the secondary compensation is tuned so that the receiver branch is purely resistive at ω (i.e., $Z_{seq}(j\omega) \approx R_L$). If the secondary is detuned or a non-negligible reactive part remains, the general expressions (3.11)–(3.18) should be used instead.

With $L_1 = L_2 = L_p = L_s = 24\,\mu\text{H}$ and $k = 0.7$, the mutual inductance is $M = kL_p = 16.8\,\mu\text{H}$, giving $\omega M = \omega k L_1$, and

$$R_L \approx \frac{(\omega k L_1)^2}{R_{\text{eq}}} \approx \frac{\left(2.26 \times 10^6 \cdot 0.7 \cdot 24 \times 10^{-6}\right)^2}{6.49} \approx 223\,\Omega. \tag{3.42}$$

Interpretation: In a rectified system, R_L is the AC equivalent of the DC load as seen at the receiver input at ω, not the physical DC load resistance.

Step 3: Shunt Capacitance
Using the canonical ZVS/ZDS relation (3.5),

$$C_1 \approx \frac{0.1836}{\omega R_{\text{eq}}} = \frac{0.1836}{(2.26 \times 10^6)(6.49)} \approx 12.5\,\text{nF}. \tag{3.43}$$

Step 4: Choose Q_L and Determine Series Elements
Selecting $Q_L = 10$. *Sensitivity note:* This bound only ensures $C_r > 0$. In practice Q_L must be large enough for the tank current to be nearly sinusoidal, so the FHA holds; values in the range $Q_L \approx 7$–10 are typical. Here $Q_L = 10$ is adopted as a conservative choice. With R_{eq} and Q_L fixed, the series inductance follows from (3.20):

$$L = \frac{Q_L R_{eq}}{\omega} \approx 28.7\,\mu\text{H}. \tag{3.44}$$

Since $L_1 = 24\,\mu\text{H}$, an external series inductance $L_{ext} = L - L_1 \approx 4.7\,\mu\text{H}$ must be added. The corresponding resonant capacitor then follows from (3.22):

$$C_r \approx \frac{1}{\omega R_{eq}\left(Q_L - \dfrac{\pi(\pi^2 - 4)}{16}\right)} \approx 7.71\,\text{nF}. \tag{3.45}$$

Step 5: Input Choke Sizing
Using (3.23),

$$L_f \approx \frac{2\left(\pi^2/4 + 1\right) R_{eq}}{f_s} \approx 125\,\mu\text{H}. \tag{3.46}$$

Note on the matching-network step. Sect. 3.1 introduces a π-type matching network for the general case in which the coupled-coil equivalent impedance is not already the canonical Class-E load. This example takes a simpler route: R_{eq} is obtained from the canonical relations, and R_L is then chosen so that, at the nominal coupling and tuning condition, it reflects to that same R_{eq}. This avoids an explicit matching network at the nominal point. If the secondary is detuned, a residual reactance remains, or the receiver-side load cannot be regulated to present the required R_L, a dedicated matching network is needed.

3.3.4 Design Summary

Table 3.2 lists the resulting first-cut component values.

3.3.5 Parameter Table and Parasitic Checklist

Table 3.2 summarizes the ideal first-cut sizing results. For simulation refinement and hardware planning, Table 3.3 lists the parasitic quantities that should be estimated

Table 3.2 First-cut component values for the design example

Quantity	Value
Optimum equivalent resistance R_{eq}	$6.49\,\Omega$
AC-equivalent secondary load target R_L (at $k = 0.7$)	$223\,\Omega$
Shunt capacitance C_1	12.5 nF
Series inductance L (total)	28.7 μH
External series inductance $L_{ext} = L - L_1$	4.7 μH
Series capacitor C_r	7.71 nF
Input choke L_f	125 μH
Peak switch voltage $V_{DS,pk}$	$\approx 3.56\,V_I \approx 53.4$ V[a]
Peak switch current $I_{S,pk}$	$\approx 2.86\,I_I \approx 3.82$ A[a]
Recommended MOSFET rating (minimum practical)	$V_{DS} \geq 80\,\text{V},\ I_D \geq 5\,\text{A}$

[a] $D = 0.5$ optimum; [1], eq. (12.42) gives the normalized peak switch current and eq. (12.43) gives the normalized peak switch voltage. Here $I_I = P_o/V_I = 1.33$ A

Table 3.3 Parasitic checklist for simulation/measurement refinement (fill with estimated or measured values)

Item	Symbol/representation	Notes (how to estimate)
Tx coil series resistance	r_1 in series with L_1	$r_1 \approx \omega L_1/Q_1$ at f_s
Rx coil series resistance	r_2 in series with L_2	$r_2 \approx \omega L_2/Q_2$ at f_s
Switch on-resistance	$r_{ds,on}$	Datasheet at temperature/current
Switch output capacitance	$C_{oss}(v)$	Use effective $C_{oss,eff}$ near operating v_{DS}
Gate drive limits	V_{GS}, R_g, deadtime	Impacts turn-on loss and ZVS margin
Layout stray inductance	L_σ (series)	Package + loop inductance (PCB)
Capacitor ESR/ESL	r_C, L_C	Especially for C_1 and C_r

from datasheets, layout calculations, and measured coil quality factors before drawing conclusions about ZVS margin, peak stress, and efficiency.

Loss estimate from the parasitic checklist. To illustrate why the checklist entries matter, consider two dominant conduction-loss terms at the design point ($V_I = 15\,\text{V}$, $P_o = 20\,\text{W}$, $f_s = 360\,\text{kHz}$).

Tx coil ESR. The tank current is approximately sinusoidal with amplitude $I_m = \sqrt{2P_o/R_{eq}} \approx 2.48$ A, giving $I_{L,rms} = I_m/\sqrt{2} \approx 1.75$ A. For a Tx coil with $Q_1 = 150$ at f_s,

$$r_1 = \frac{\omega L_1}{Q_1} = \frac{2\pi \times 360 \times 10^3 \times 24 \times 10^{-6}}{150} \approx 0.36\,\Omega, \tag{3.47}$$

so the estimated Tx-coil conduction loss is

$$P_{coil} = I_{L,rms}^2\, r_1 \approx (1.75)^2 \times 0.36 \approx 1.1\,\text{W},$$

which is about 5.6% of P_o.

Switch $r_{\text{ds,on}}$. For the $D = 0.5$ optimum waveform, the switch RMS current is $I_{S,\text{rms}} \approx 1.54\, I_I \approx 2.05\,\text{A}$, where $I_I = P_o/V_I = 1.33\,\text{A}$ and the factor 1.54 follows from the optimum Class-E switch-current waveform [1]. With a representative $r_{\text{ds,on}} = 150\,\text{m}\Omega$ for an 80 V/5 A MOSFET,

$$P_{\text{sw}} = I_{S,\text{rms}}^2\, r_{\text{ds,on}} \approx (2.05)^2 \times 0.15 \approx 0.63\,\text{W},$$

which is about 3.2% of P_o.

The combined conduction loss is therefore about 1.75 W, giving a conduction-only efficiency of roughly 92% before Rx-coil ESR, capacitor ESR, and switching losses are added. The Rx coil ESR (r_2), capacitor ESR (r_C), and C_{oss}-related switching loss are additive and will further reduce efficiency. Their magnitudes depend on component selection and should be estimated from datasheets and coil measurements before any efficiency budget is finalised. Table 3.3 is there to ensure no significant loss term is missed before building a prototype.

Sensitivity to Coupling Variation

Circuit-based simulations of this design over a range of coupling coefficients show that ZVS is achieved at the nominal point $k = 0.7$. As k increases toward 0.8, some hard switching appears but remains mild; at the same time the resonant current amplitude falls, reducing conduction losses, so overall efficiency stays above 90%. Below the nominal k, the antiparallel diode begins to conduct (lack of ZDS) and efficiency falls more sharply, dropping to the upper 70% range. The response is therefore asymmetric: the design tolerates moderate over-coupling better than under-coupling, which is relevant when choosing the nominal design point. In practical Qi2 systems, the coupling coefficient is typically in the 0.3–0.6 range, so the present example, with $k = 0.7$, represents an optimistic nominal condition; a more conservative design point would shift the ZVS threshold and the efficiency contour accordingly.

References

1. M.K. Kazimierczuk, D. Czarkowski, *Resonant Power Converters*, 2nd ed. (Wiley, Hoboken, 2011). ISBN: 978-0-470-90538-8
2. Y.-H. Wu, Wireless communication system for wireless charging of electric vehicle. Ph.D. dissertation, New York University Tandon School of Engineering (May 2020)
3. A. Ayachit, F. Corti, A. Reatti, M.K. Kazimierczuk, Zero-voltage switching operation of transformer Class-E inverter at any coupling coefficient. IEEE Trans. Ind. Electron. **66**(3), 1809–1819 (2019). https://doi.org/10.1109/tie.2018.2838059
4. F. Corti, A. Reatti, Y.-H. Wu, D. Czarkowski, S. Musumeci, Zero voltage switching condition in Class-E inverter for capacitive wireless power transfer applications. Energies **14**(4), 911 (2021). https://doi.org/10.3390/en14040911
5. B.C. Nicoletti, F.E. Bisogno, L.S. Mendonça, Exact solution of the Class-E inverter for wireless power transfer systems. Eletrônica de Potência **30**, e202542 (2025). https://doi.org/10.18618/rep.e202542

6. L. Roslaniec, A.S. Jurkov, A. Al Bastami, D.J. Perreault, Design of single-switch inverters for variable resistance/load modulation operation. IEEE Trans. Power Electron. **30**(6), 3200–3214 (2015). https://doi.org/10.1109/tpel.2014.2331494
7. A. Komanaka, W. Zhu, X. Wei, K. Nguyen, H. Sekiya, Generalized analysis of load-independent ZCS parallel-resonant inverter. IEEE Trans. Ind. Electron. **69**(1), 347–356 (2022). https://doi.org/10.1109/tie.2021.3053888
8. W. Zhu, Y. Komiyama, A. Komanaka, K. Nguyen, H. Sekiya, Analysis of load-independent ZCS parallel-resonant inverter with constant current. IEEE Trans. Ind. Electron. **71**(9), 10433–10444 (2024). https://doi.org/10.1109/tie.2023.3333019
9. Y. Komiyama, W. Zhu, A. Konishi, K. Nguyen, H. Sekiya, Analysis and design of load-independent Class-E zero-voltage switching power oscillator. IEEE Trans. Circuits Syst. I: Regular Pap. **73**(1), 721–734 (2026). https://doi.org/10.1109/tcsi.2025.3594362
10. H. Sekiya, K. Tokano, W. Zhu, Y. Komiyama, K. Nguyen, Design procedure of load-independent Class-E WPT systems and its application in robot arm. IEEE Trans. Ind. Electron. **70**(10), 10014–10023 (2023). https://doi.org/10.1109/tie.2022.3220818
11. L. Albertoni, F. Grasso, J. Matteucci, M.C. Piccirilli, A. Reatti, A. Ayachit, M.K. Kazimierczuk, Analysis and design of full-bridge Class-DE inverter at fixed duty cycle, in *IECON 2016–42nd Annual Conference of the IEEE Industrial Electronics Society* (2016), pp. 5609–5614. https://doi.org/10.1109/iecon.2016.7793240
12. H. Koizumi, T. Suetsugu, M. Fujii, K. Shinoda, S. Mori, K. Ikeda, Class DE high-efficiency tuned power amplifier. IEEE Trans. Circuits Syst. I: Fundam. Theory Appl. **43**(1), 51–60 (1996). https://doi.org/10.1109/81.481461
13. D.C. Hamill, Class DE inverters and rectifiers for DC–DC conversion, in *Proceedings of the 27th Annual IEEE Power Electronics Specialists Conference (PESC)* (1996), pp. 854–860. https://doi.org/10.1109/pesc.1996.548681
14. K. Inoue, T. Nagashima, X. Wei, H. Sekiya, Design of high-efficiency inductive-coupled wireless power transfer system with Class-DE transmitter and Class-E rectifier, in *IECON 2013 – 39th Annual Conference of the IEEE Industrial Electronics Society* (2013), pp. 613–618. https://doi.org/10.1109/iecon.2013.6699205
15. T. Nagashima, X. Wei, H. Sekiya, Analytical design procedure for resonant inductively coupled wireless power transfer system with Class-DE inverter and Class-E rectifier, in *2014 IEEE Asia Pacific Conference on Circuits and Systems (APCCAS)* (2014), pp. 288–291. https://doi.org/10.1109/apccas.2014.7032776
16. H. Sarnago, Ò. Lucía, A. Mediano, J.M. Burdío, Class-D/DE dual-mode-operation resonant converter for improved-efficiency domestic induction heating system. IEEE Trans. Power Electron. **28**(3), 1274–1285 (2013). https://doi.org/10.1109/tpel.2012.2206405
17. H. Sekiya, X. Wei, T. Nagashima, M.K. Kazimierczuk, Steady-state analysis and design of Class-DE inverter at any duty ratio. IEEE Trans. Power Electron. **30**(7), 3685–3694 (2015). https://doi.org/10.1109/tpel.2014.2339355
18. H. Sekiya, S. Nemoto, J. Lu, T. Yahagi, Phase control for resonant DC–DC converter with Class-DE inverter and Class-E rectifier. IEEE Trans. Circuits Syst. I: Regul. Pap. **53**(2), 254–263 (2006). https://doi.org/10.1109/tcsi.2005.855741
19. P. Wang, Q. Li, Y. Liu, W. Yuan, K. Yan, Z. Pang, A novel impedance matching of Class DE inverter for high efficiency, wide-impedance WPT systems. Electronics **13**(5), 959 (2024). https://doi.org/10.3390/electronics13050959
20. T. Nagashima, X. Wei, E. Bou-Balust, E. Alarcón, H. Sekiya, Analytical design for resonant inductive-coupling wireless power transfer system with Class-E inverter and Class-DE rectifier, in *2015 IEEE International Symposium on Circuits and Systems (ISCAS)* (2015), pp. 686–689. https://doi.org/10.1109/iscas.2015.7168726
21. D. Kawamoto, H. Sekiya, H. Koizumi, I. Sasase, S. Mori, Design of phase-controlled Class E inverter with asymmetric circuit configuration. IEEE Trans. Circuits Syst. II: Express Briefs **51**(10), 523–527 (2004). https://doi.org/10.1109/tcsii.2004.834546

$$
\begin{aligned}
Z_{\text{ref}} &= \frac{(\omega M)^2}{Z_2} \\
&= \frac{\left((560.01 \times 10^3)(33 \times 10^{-6} \times 0.2)\right)^2}{5.05 + j1.6723} \\
&= 2.438 - j0.807
\end{aligned} \tag{2.44}
$$

Using Eq. (2.11), the input impedance angle is

$$
\begin{aligned}
\psi &= \arctan\left(\frac{\text{imag}(Z_1 + Z_{\text{ref}})}{\text{real}(Z_1 + Z_{\text{ref}})}\right) \\
&= \arctan\left(\frac{\text{imag}(0.15 + j1.6723 + 2.438 - j0.807)}{\text{real}(0.15 + j1.6723 + 2.438 - j0.807)}\right) \\
&= \arctan\left(\frac{0.8653}{2.588}\right) = 18.5^\circ
\end{aligned} \tag{2.45}
$$

Using Eq. (2.18), and considering that the amplitude of the fundamental component of the full-bridge inverter output voltage is twice that of the half-bridge case, the received power is

$$
\begin{aligned}
P_2 &= \left(\frac{0.9\omega M V_I}{|Z_2|\,|Z_1 + Z_{\text{ref}}|}\right)^2 R_L \\
&= \left(\frac{0.9(560.01 \times 10^3)(6.6 \times 10^{-6})(400)}{|5.05 + j1.6723|\,|0.15 + j1.6723 + 2.438 - j0.807|}\right)^2 \times 5 \\
&= 42.01\,\text{kW}
\end{aligned} \tag{2.46}
$$

Using Eq. (2.24), the efficiency is

$$
\begin{aligned}
\eta &= \frac{\text{real}(Z_{\text{ref}})}{\text{real}(Z_1 + Z_{\text{ref}})} \cdot \frac{R_L}{\text{real}(Z_2)} \\
&= \frac{\text{real}(2.438 - j0.807)}{\text{real}(0.15 + j1.6723 + 2.438 - j0.807)} \cdot \frac{5}{\text{real}(5.05 + j1.6723)} \\
&= 93.27\%
\end{aligned} \tag{2.47}
$$

Comparison of Examples 2.1 and 2.2 shows that the WPT system with the full-bridge inverter has a higher power-output capability, but a slightly lower efficiency.

Example 2.3 Evaluate the efficiency when the inverter switching loss is considered under the parameters of Example 2.1. Assume that the MOSFET IPW65R041CFD is employed. The rising time and falling time of the MOSFET are 28 ns and 8 ns, respectively.

22. F. Issi, O. Kaplan, Design and application of wireless power transfer using Class-E inverter based on adaptive impedance-matching network. ISA Trans. **126**, 415–427 (2022). https://doi.org/10.1016/j.isatra.2021.07.050
23. F. Issi, Modeling and analysis of a Class-E bidirectional wireless power transfer system with phase-shift control, in *2025 IEEE 19th International Conference on Compatibility, Power Electronics and Power Engineering (CPE-POWERENG)* (2025). https://doi.org/10.1109/cpe-powereng63314.2025.11027208
24. C.-M. Lai, D.-T. Lin, H.-E. Liu, X.F. Chen, C.K. Tse, An interleaved Class-E resonant inverter with wide-range load for a single-stage wireless DC motor drive, in *2024 IEEE Wireless Power Technology Conference and Expo (WPTCE)* (2024), pp. 275–279. https://doi.org/10.1109/wptce59894.2024.10557429
25. I. Nikiforidis, K. Bampouras, P. Wagle, D.C. Yates, P.D. Mitcheson, A 9 kW, 3.47 MHz wireless power transfer system with a parallel differential Class E inverter for industrial applications. IEEE J. Emerg. Sel. Top. Power Electron. **13**(4), 4171–4190 (2025). https://doi.org/10.1109/jestpe.2024.3497712
26. Wireless Power Consortium, Qi wireless charging. Accessed 20 Mar 2026 [Online]. Available: https://www.wirelesspowerconsortium.com/standards/qi-wireless-charging/
27. A. Munsi, R. Kumar, K. Aditya, Hardware design and implementation for Class-E power amplifier for Qi compatible wireless power transfer for consumer electronics. Int. J. Power Electron. **21**(4), 398–415 (2025)

Chapter 4
Receiver Side Converters

This chapter develops analytical and design-oriented frameworks for the receiver-side power conversion and interface circuits of wireless power transfer (WPT) systems, covering impedance matching networks, rectifiers, and DC–DC converters. First, impedance matching techniques, including inductive and capacitive tuning, are examined to ensure optimal coupling and maximum power transfer. The chapter then presents detailed analyses of rectifier topologies, including Class D current-driven and voltage-driven rectifiers, as well as semi-bridgeless configurations, with emphasis on conduction behavior, waveform characteristics, and voltage conversion relationships. Finally, DC–DC converter topologies—step-up, step-down, and buck-boost—are discussed in the context of voltage regulation and system-level integration. The methodology combines theoretical modeling and circuit analysis to derive key performance metrics, providing practical insights for the design and optimization of high-efficiency WPT systems.

4.1 Impedance Matching Circuits

4.1.1 Inductive Tuning

Most real-time methods track the movement of resonance point and tune the circuit by adjusting operating frequency or variable reactance. This section introduces a method for inductance control based on a Class E rectifier for wireless power transfer systems. The method controls the secondary inductance of the wireless power transfer system with a Class E full bridge rectifier and PWM. [1–3]

The circuit of a Class E full-bridge low di/dt rectifier with a series inductor is shown in Fig. 4.1a. The rectifier consists of four rectifying diodes and an inductor

D. Czarkowski et al., *Converter Topologies for Resonant Wireless Power Transfer*, Power Systems, https://doi.org/10.1007/978-3-032-31448-2_4

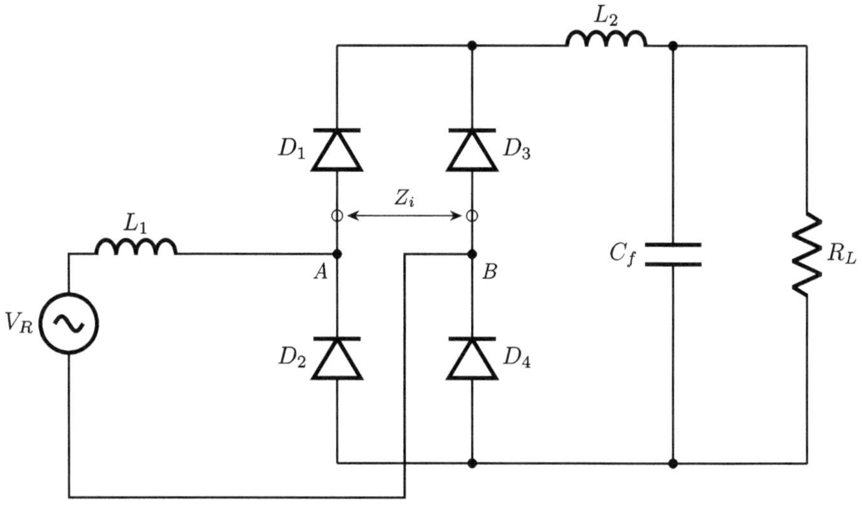

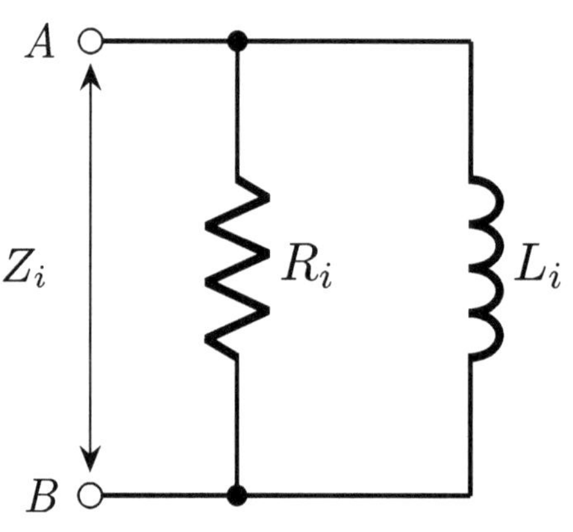

Fig. 4.1 (**a**) Full-bridge Class E low di/dt rectifier with a series inductor; nodes A and B mark the bridge AC input terminals and Z_i denotes the impedance seen at those terminals. (**b**) Fundamental-frequency equivalent of Z_i, represented by the parallel combination of R_i and L_i

L_2 connected in series with the parallel combination of a capacitor C_f and load resistance R_L, which form a first-order low-pass output filter.

The rectifier is driven by a sinusoidal voltage source V_R, which delivers dc power to the resistive load R_L. An inductance L_1 is connected in series with V_R; it corresponds to the inductance of the coil of the secondary circuit of a WPT system.

The idealized voltage and current waveforms in the circuit of Fig. 4.1a are presented in Fig. 4.2. As can be observed, the input voltage v_R is a sine wave and the output voltage V_O is constant. The rectifier diodes D_1 and D_2 are turned on due to commutation in $0 < \omega t < \pi D$ and $\pi < \omega t < \pi + \pi D$ time intervals, respectively, where D is the duty ratio. Hence, v_{AB} is equal to zero during these intervals.

Noting that the currents through the inductors change slowly, v_R and v_{AB} can be described as follows:

$$v_R = V_m \sin \omega t \tag{4.1}$$

$$v_{AB} = \begin{cases} 0, & 0 < \omega t < \pi D, \\ V_m \sin \omega t, & \pi D < \omega t < \pi \end{cases} \tag{4.2}$$

In the steady state, V_O is equal to the average value of v_{AB} in a cycle and $v_{L2} = v_{AB} - V_O$, thus:

$$V_O = \frac{1}{\pi} \int_0^{\pi} V_{AB} d\omega t = \frac{1}{\pi} \int_{\pi D}^{\pi} V_{AB} d\omega t = \frac{V_m}{\pi} (\cos \pi D + 1) \tag{4.3}$$

$$v_{L2} = \begin{cases} -\dfrac{V_m}{\pi} (\cos \pi D + 1), & 0 < \omega t < \pi D, \\ V_m \left(\sin \omega t - \dfrac{\cos \pi D + 1}{\pi} \right), & \pi D < \omega t < \pi \end{cases} \tag{4.4}$$

$$v_{L1} = \begin{cases} V_m \sin \omega t & , \ 0 < \omega t < \pi D \\ V_m \left(\sin \omega t - \dfrac{\cos \pi D + 1}{\pi} \right), & \pi D < \omega t < \pi \end{cases} \tag{4.5}$$

From the connections of the inductors and the diodes, the relationships among the currents are as follows:

$$i_{D1} = \frac{1}{2} (i_{L2} + i_{L1}) \tag{4.6}$$

$$i_{D2} = \frac{1}{2} (i_{L2} - i_{L1}) \tag{4.7}$$

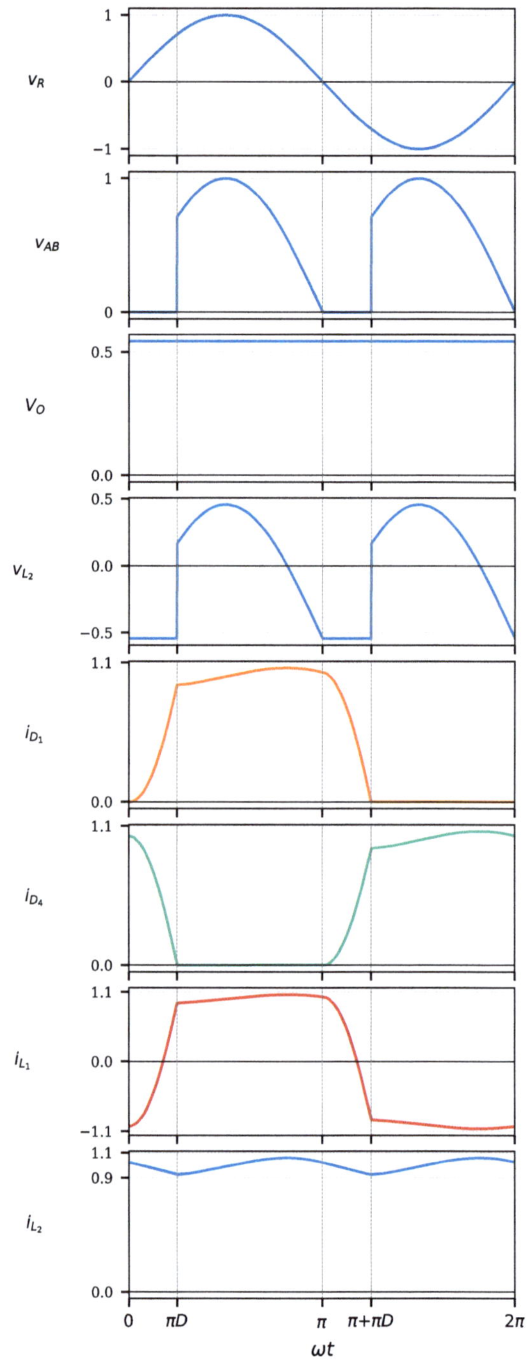

Fig. 4.2 Ideal voltage and current waveforms of a full-bridge Class E rectifier

With (4.4), (4.5), (4.6), (4.7) and the conditions for steady-state i_{D1} and i_{D2}, i_{L1} and i_{L2} are expressed as follows:

$$i_{L1} = \begin{cases} -\dfrac{V_m}{\omega L_1}\cos\omega t - \dfrac{V_m}{2}\left(\dfrac{D\cos\pi D + D}{\omega L_2} - \dfrac{1+\cos\pi D}{\omega L_1}\right), \\ \qquad 0 < \omega t < \pi D \\ \dfrac{V_m}{2\omega L_1}(1-\cos\pi D) \\ \quad -\dfrac{V_m}{\omega L_2}\left(\cos\omega t + \dfrac{\cos\pi D + 1}{\pi}\omega t - \cos\pi D - \dfrac{D\cos\pi D + D}{2}\right), \\ \qquad \pi D < \omega t < \pi \end{cases} \tag{4.8}$$

$$i_{L2} = \begin{cases} -\dfrac{V_m}{\pi\omega L_2}(\cos\pi D + 1)\,\omega t + \dfrac{V_m}{2}\left(\dfrac{D\cos\pi D + D}{\omega L_2} + \dfrac{1-\cos\pi D}{\omega L_1}\right), \\ \qquad 0 < \omega t < \pi D \\ \dfrac{V_m}{2\omega L_1}(1-\cos\pi D) \\ \quad -\dfrac{V_m}{\omega L_2}\left(\cos\omega t + \dfrac{\cos\pi D + 1}{\pi}\omega t - \cos\pi D - \dfrac{D\cos\pi D + D}{2}\right), \\ \qquad \pi D < \omega t < \pi \end{cases} \tag{4.9}$$

Since the average current through C_f in steady-state is zero, the dc component of output current I_O is the average of i_{L2} in (4.9).

$$\begin{aligned} I_O &= \frac{1}{\pi}\int_0^{\pi} i_{L2}\, d\omega t \\ &= \frac{\pi I_O R_L}{\cos\pi D + 1}\left[(D-1)\frac{1-\cos\pi D}{2\omega L_2} + \frac{\sin\pi D}{\pi\omega L_2} + \frac{1-\cos\pi D}{2\omega L_1}\right] \end{aligned} \tag{4.10}$$

Thus, the relationship between the duty ratio D and the circuit components is

$$R_L = \frac{\cos\pi D + 1}{\pi(D-1)\dfrac{1-\cos\pi D}{2\omega L_2} + \dfrac{\sin\pi D}{\omega L_2} + \pi\dfrac{1-\cos\pi D}{2\omega L_1}} \tag{4.11}$$

Nodes A and B in Fig. 4.1a mark the AC input terminals of the bridge; v_{AB} denotes the voltage at A with respect to B. Since v_R is sinusoidal, the input impedance Z_i seen at terminals A–B can be depicted as a parallel combination of the input resistance R_i and the input inductance L_i, as shown in Fig. 4.1b. The fundamental component of the input current is

$$i_i = i_{Ri} + i_{Li} = I_{Rim} \sin \omega t - I_{Lim} \cos \omega t \quad (4.12)$$

where I_{Rim} and I_{Lim} are the amplitudes of the currents through R_i and L_i. If power loss is neglected, the input power equals the output power and the input resistance R_i is described as follows:

$$\frac{V_m^2}{2R_i} = \frac{V_O^2}{R_L} \quad (4.13)$$

$$R_i = \frac{1}{2}\left(\frac{V_m}{V_O}\right)^2 R_L = \frac{1}{2}\left(\frac{\pi}{\cos \pi D + 1}\right)^2 R_L \quad (4.14)$$

The input inductance L_i is derived from the Fourier series formula of the input current i_{L1} by using (4.8) as follows:

$$\begin{aligned} I_{Lim} &= -\frac{1}{\pi}\int_0^{2\pi} i_{L1} \cos \omega t \, d\omega t \\ &= \frac{V_m}{\omega L_2}\left(\frac{\sin \pi D \cos \pi D}{\pi} - \frac{2(\cos \pi D + 1)^2}{\pi^2} + 1 - D\right) \\ &\quad + \frac{V_m}{\omega L_1}\left(D - \frac{\sin \pi D \cos \pi D}{\pi}\right) \end{aligned} \quad (4.15)$$

$$\begin{aligned} L_i &= \frac{V_m}{\omega I_{Lim}} \\ &= \left[\frac{\frac{\sin \pi D \cos \pi D}{\pi} - \frac{2(\cos \pi D + 1)^2}{\pi^2} + 1 - D}{L_2} + \frac{D - \frac{\sin \pi D \cos \pi D}{\pi}}{L_1}\right]^{-1} \end{aligned} \quad (4.16)$$

Equation (4.16) presents the input inductance L_i at the fundamental frequency f. As can be observed, L_i is a function of the inductances L_1, L_2 and the duty ratio D, which in turn depends on the values of L_1, L_2 and R_L according to (4.11). As a result, L_i is a function of the inductances L_1, L_2 and the load R_L. Accordingly, L_i can be controlled by controlling R_L.

The analysis above assumes that L_2 is sufficiently large to maintain a nearly constant current i_{L2}, so that the piecewise-linear waveforms of (4.8)–(4.9) and the

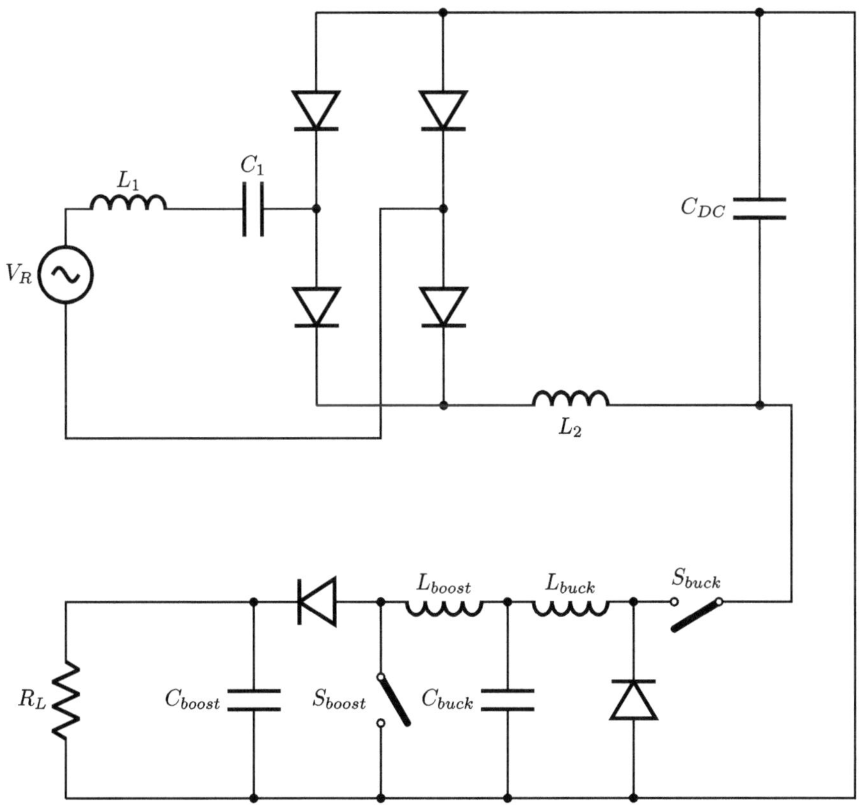

Fig. 4.3 Secondary pick-up comprising secondary coil L_1, resonant capacitor C_1, Class E full-bridge rectifier, series filter inductor L_2, dc link capacitor C_{DC}, and the cascaded buck–boost converter with load R_L

closed-form expressions (4.11) and (4.16) are valid. In the complete secondary pick-up circuit of Fig. 4.3, a resonant capacitor C_1 is added in series with L_1 and a smaller value of L_2 is used, which invalidates these piecewise waveforms. The equivalent inductance L_i and the tuning relationship (4.16) are still used as design quantities at the fundamental frequency, but the actual circuit waveforms are obtained by simulation.

The topology of the secondary pick-up comprises a secondary coil L_1, a resonant capacitor C_1, a Class E full-bridge rectifier, two DC–DC converters in cascade and a load resistance R_L as shown in Fig. 4.3.

The capacitance C_1 is determined to compensate the inductance L_1 of the coil at the operating frequency f by

$$C_1 = \frac{1}{\omega^2 L_1} \tag{4.17}$$

where $\omega = 2\pi f$. Because C_1 is determined at the design stage, the circuit is vulnerable to inductance variations of L_1 during operation. In practice, the rectifier input impedance Z_i presents a small residual inductive component at the nominal operating point even after D_1 and D_2 are set, so the series L_1–C_1 branch must be made slightly capacitive to achieve net secondary resonance. This requires choosing C_1 somewhat *smaller* than the value given by (4.17), so that $1/(\omega C_1) > \omega L_1$ and the branch presents a net capacitive reactance that cancels the rectifier's inductive contribution. In the simulation that follows, $C_1 = 33\,\text{nF}$ is used instead of the theoretical 35.06 nF; the smaller value shifts the L_1–C_1 branch to the capacitive side and simplifies the initial selection of D_1 and D_2. The value of C_1 does not appear in (4.20), so that equation remains valid regardless of this practical adjustment.

The DC–DC buck converter is followed by a DC–DC boost converter, which control the input resistance through the duty ratios D_1 and D_2 for the buck and boost converters, respectively.

The ideal buck and boost converters have the input resistance and output voltage as follows:

$$R_{in,buck} = \frac{V_{i,buck}}{I_{i,buck}} = \frac{1}{D_1^2} R_{in,boost},$$
$$R_{in,boost} = \left(\frac{1 - D_2}{D_1}\right)^2 R_L \tag{4.18}$$

$$\frac{V_{o,boost}}{V_{i,buck}} = \frac{D_1}{1 - D_2} \tag{4.19}$$

where $R_{in,buck}$ also serves as the load of the Class E rectifier. Substituting (4.18) into (4.11), the relationship between the DC–DC converter duty ratios D_1, D_2 and the diode on-duty cycle D is obtained as follows:

$$\left(\frac{1 - D_2}{D_1}\right)^2 R_L = \frac{\cos\pi D + 1}{\pi(D-1)\dfrac{1-\cos\pi D}{2\omega L_2} + \dfrac{\sin\pi D}{\omega L_2} + \pi\dfrac{1-\cos\pi D}{2\omega L_1}} \tag{4.20}$$

where R_L is the load of the whole circuit. The family of curves relating D to D_1 for representative values of D_2 is shown in Fig. 4.4.

The circuit is tuned by adjusting the duty ratios D_1 and D_2, which in turn change $R_{in,buck}$, which corresponds to the load of the Class E rectifier. According to Eq. (4.16), once the load of the Class E rectifier is determined, the equivalent inductance is also determined. Note that the inductance L_i in Eq. (4.16) does not indicate the input impedance of the Class E rectifier but the equivalent inductance of the secondary side.

Since every resonant circuit has an equivalent inductance L_i that is susceptible to component variations, the circuit should be able to respond quickly to variations in L_i within an appropriate range. Equation (4.16) also shows that inductor L_2 affects

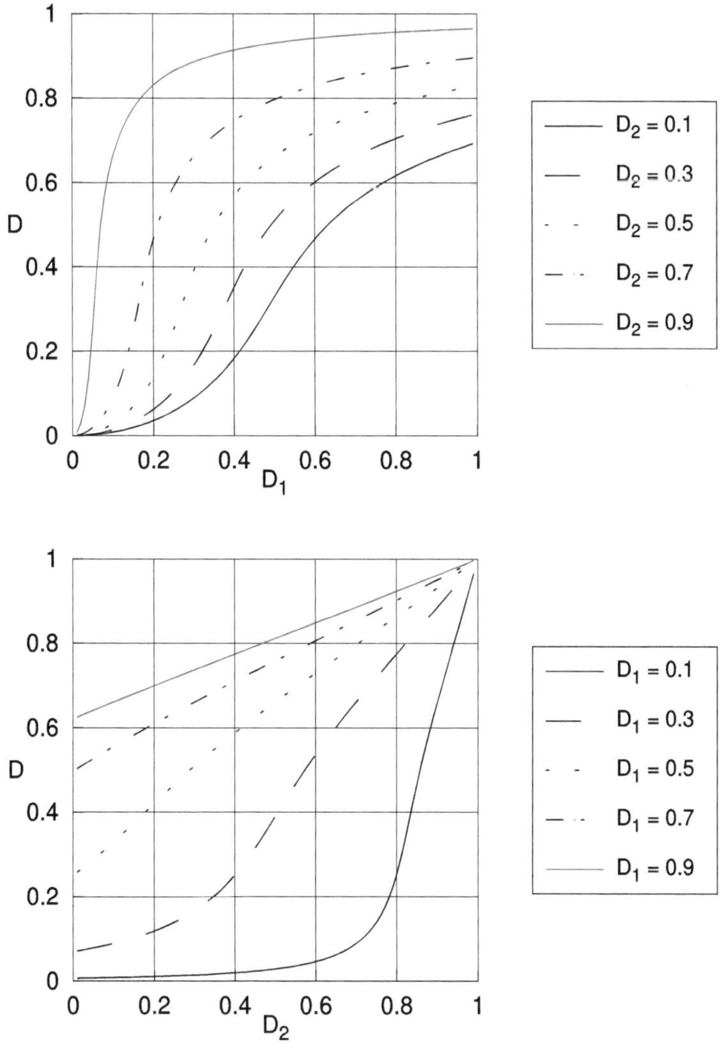

Fig. 4.4 Diode on-duty cycle D as a function of D_1 for $D_2 = 0.1,\ 0.3,\ 0.5,\ 0.7,\ 0.9$ (from top to bottom), computed from (4.20) with $L_1 = 100\,\mu\text{H}$, $L_2 = 20\,\mu\text{H}$, $R_L = 10\,\Omega$, $f = 85\,\text{kHz}$

the value of L_i. Figure 4.5 illustrates the variation of the equivalent inductance L_i versus D for different values of L_2. For better tuning effect, L_i is supposed to satisfy both: response speed and range. For less switching stress and loss, the diode duty cycle D should vary within the middle range. In conclusion, L_2 must avoid being either too large or too small relative to L_1.

From Fig. 4.5 and after analyzing multiple parameter sets, a practical range for L_2 of about $1/5$ to 1 of L_1 is recommended.

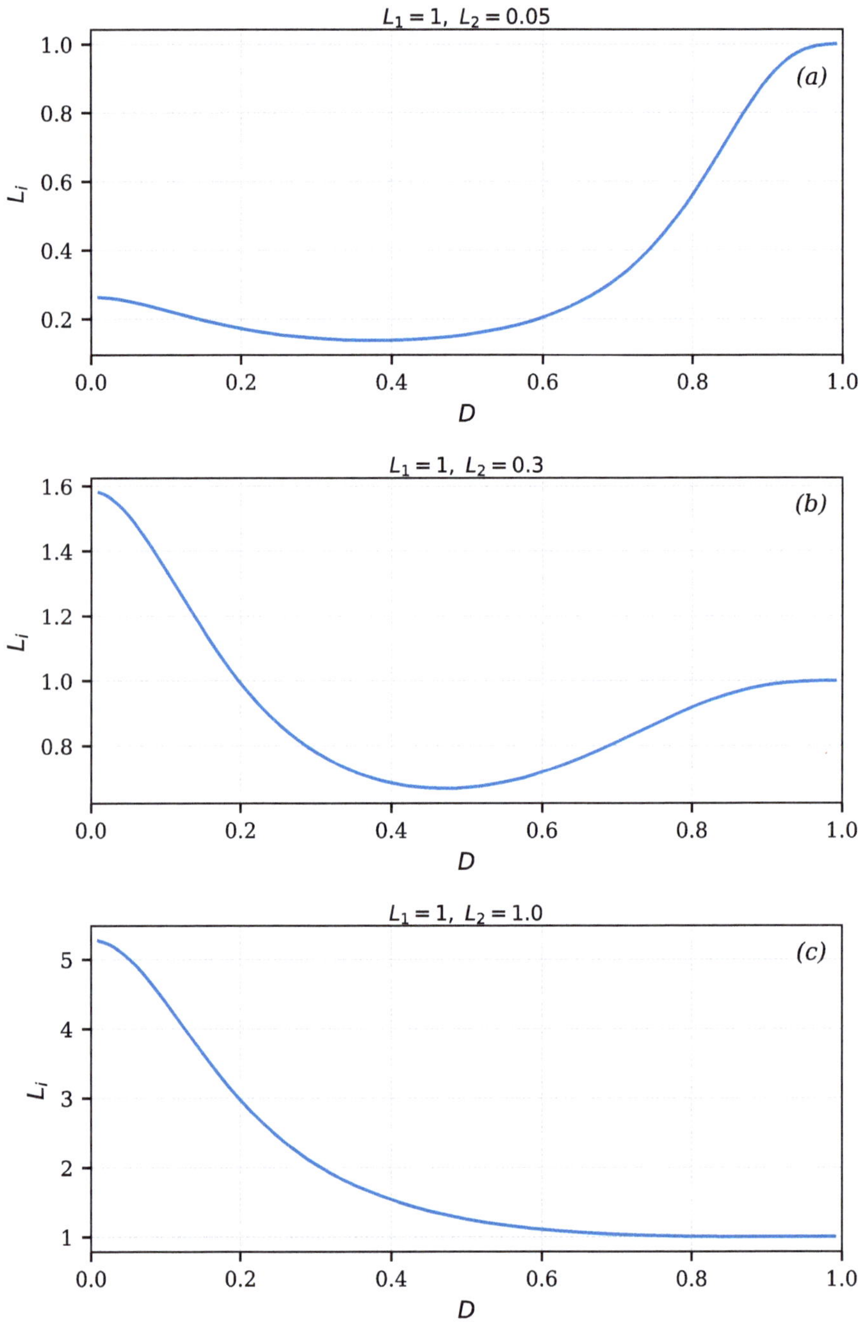

Fig. 4.5 L_i versus D at different L_2: (**a**) $L_1 = 1$, $L_2 = 0.05$; (**b**) $L_1 = 1$, $L_2 = 0.3$; (**c**) $L_1 = 1$, $L_2 = 1$

It is also noted that each diode on-duty cycle D corresponds to many combinations of D_1 and D_2 that generate different output voltages according to (4.19). Consequently, it is necessary to carefully select a proper combination of D_1 and D_2 in order to keep the circuit in resonance and also to make the output voltage as close as possible to the required value.

For subsequent discussion it is convenient to define

$$q = \frac{1 - D_2}{D_1}. \tag{4.21}$$

Then (4.20) may be written as $q^2 R_L = f(D)$, where $f(D)$ denotes the right-hand side of (4.20). Consequently, once exact retuning is imposed for a fixed load R_L, the required diode conduction interval D determines a unique value of q. Under the ideal buck–boost relations of (4.18) and (4.19), the overall dc-side voltage conversion ratio (where $V_p \equiv V_{i,buck}$ denotes the voltage at the buck input) is

$$\frac{V_O}{V_p} = \frac{D_1}{1 - D_2} = \frac{1}{q}. \tag{4.22}$$

Therefore, the many duty-ratio pairs (D_1, D_2) that satisfy the same exactly retuned value of q do not provide an independent dc voltage-control degree of freedom when R_L is fixed. Instead, they mainly redistribute current ripple, semiconductor stress, and internal converter operating margins between the buck and boost stages.

According to the previous analysis, a set of component values is selected as follows: $L_1 = 100\,\mu\text{H}$, $L_2 = 20\,\mu\text{H}$, $C_1 = 33\,\text{nF}$, $R_L = 10\,\Omega$, $V_{R,\text{peak}} = 50\,\text{V}$, and $f = 85\,\text{kHz}$. The circuit is tuned to resonate with $D_1 = 0.70$ and $D_2 = 0.57$. The following results are obtained by circuit simulation of the complete topology of Fig. 4.3.

Figure 4.6 shows the waveforms of input voltage v_R and inductor current i_{L1} under three operating conditions. As can be observed from Fig. 4.6a, the input voltage and current are approximately in phase, confirming that the circuit is resonating.

If L_1 is decreased by 20%, that is, to $80\,\mu\text{H}$, while C_1 remains fixed at 33 nF, the circuit becomes detuned. Figure 4.6b clearly shows that the circuit is no longer resonating: the current leads the voltage, reflecting the capacitive character of the secondary seen at the source terminals when L_1 falls below its design value. Tuning is then recovered by adjusting D_1 and D_2 to 0.30 and 0.40 respectively, and the corresponding waveforms are shown in Fig. 4.6c.

The tuning process operates as follows: by adjusting D_1 and D_2 a new value for $R_{in,buck}$ is established, which in turn modifies the diode duty cycle D and consequently the equivalent inductance L_i. As a result, the output voltage increases while I_{L1} decreases compared to the untuned case, corresponding to an improvement in efficiency and power transferred to the load.

It should be noted that while i_{L1} appears nearly sinusoidal in Fig. 4.6a and b, the retuned waveform in panel (c) exhibits measurable harmonic distortion. This arises

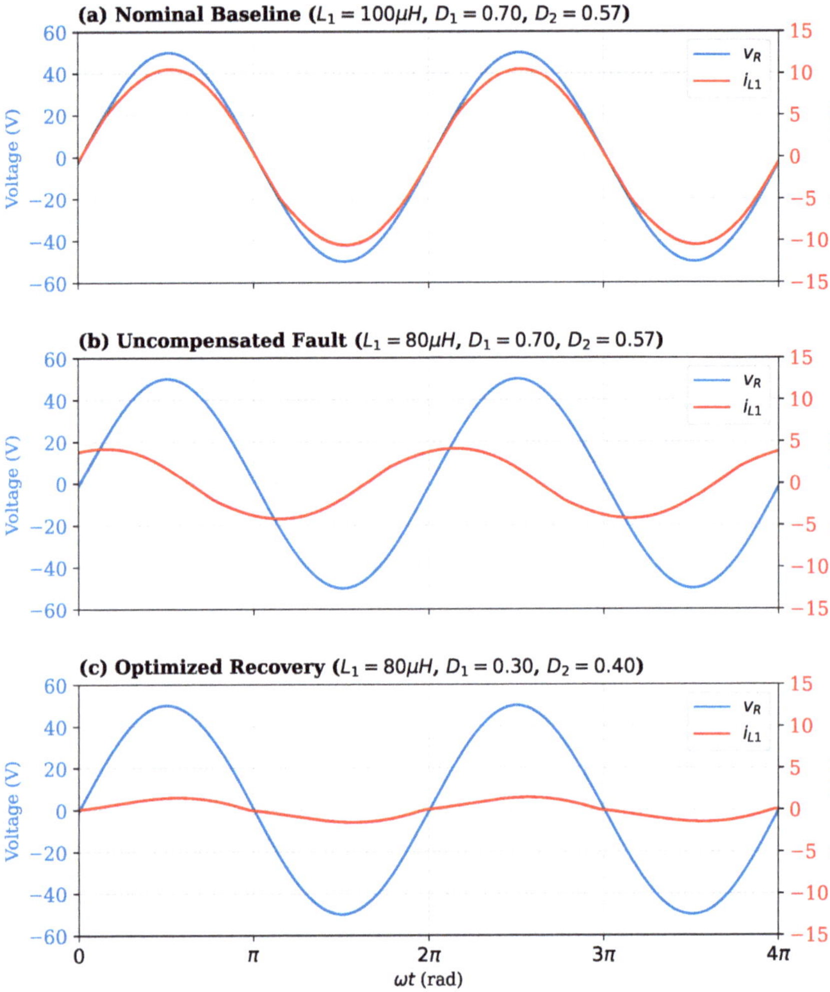

Fig. 4.6 Waveforms of input voltage v_R and inductor current i_{L1}: (**a**) nominal baseline, $L_1 = 100\,\mu\text{H}$, $D_1 = 0.70$, $D_2 = 0.57$; (**b**) uncompensated fault, $L_1 = 80\,\mu\text{H}$, $D_1 = 0.70$, $D_2 = 0.57$; (**c**) optimized recovery, $L_1 = 80\,\mu\text{H}$, $D_1 = 0.30$, $D_2 = 0.40$

because retuning increases the reflected input resistance R_i of the rectifier, which reduces the loaded quality factor of the input resonant branch. The fixed current scale used across all three panels, chosen to facilitate direct amplitude comparison, visually suppresses this distortion in panel (c).

4.1.2 Capacitive Tuning

In a wireless power transfer (WPT) system via the magnetic resonant coupling, one of the most challenging design issues is to maintain a reasonable level of power transfer efficiency (PTE), even when the distance between the transmitter and the receiver changes. When the distance varies, the PTE drastically decreases due to the impedance mismatch between the resonator of the transmitter and that of the receiver [3, 4]. This section will present a novel series/parallel capacitor matrix in the transmitter, where the impedance can be automatically reconfigured to track the optimum impedance-matching point in the case of varying distances. The dynamic WPT matching system is enabled by changing the combination of series and parallel capacitors in the capacitor matrix [5].

A block diagram of a WPT transceiver system that contains an impedance matching mechanism is shown in Fig. 4.7. The transmitter is composed of a power amplifier, matching network, and antenna. The receiver is composed of a receiving antenna, matching network, rectifier, DC–DC converter, and load.

During WPT with magnetic resonators (antennas), when the impedances of the transmitter and receiver are not perfectly matched, the system never reaches the achievable maximum PTE. Simulation results illustrating power transmission loss due to distance variation are presented in this section. The simulations were performed using Agilent Advanced Design System (ADS). The source coil geometry is illustrated in Fig. 4.8, and the simulation parameters are listed in Table 4.1. The source and load coils have the same parameters.

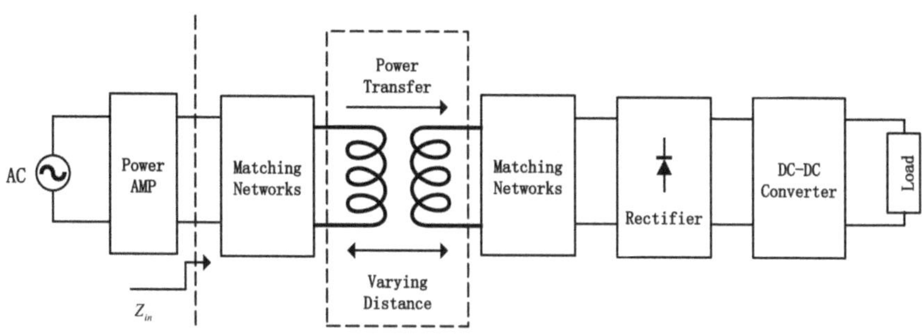

Fig. 4.7 Block diagram of a WPT transceiver system

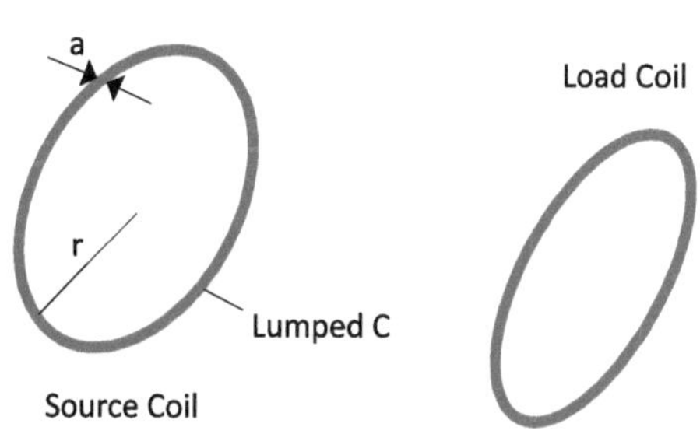

Fig. 4.8 Source coil geometry used in the ADS simulation setup, showing coil radius r, wire radius a, and lumped capacitance C

Table 4.1 WPT simulation parameters. The source and load coils have the same parameters

r (m)	a (mm)	L (μH)	Quality factor	C (pF)	R (Ω)	Freq. (MHz)
0.2	5.8	1.084	1354	127	0.068	13.56

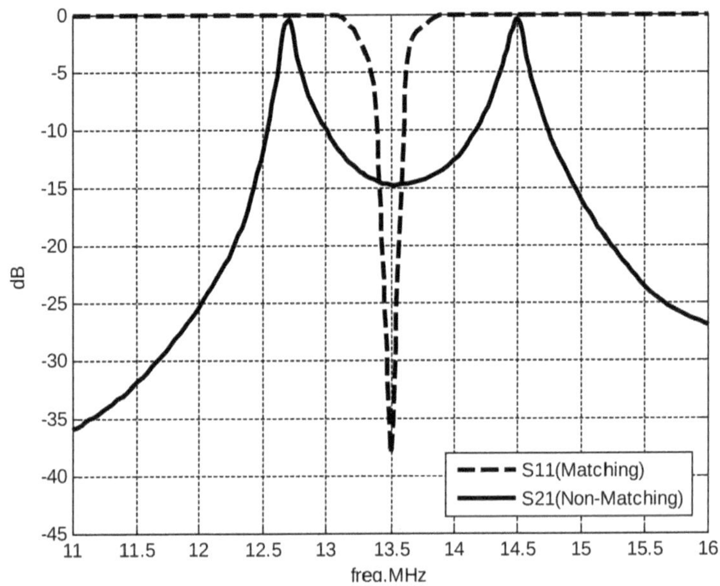

Fig. 4.9 S-parameter simulations with matching impedance and mismatching cases

Initially, with a distance of 0.6 m between the transmitter and the receiver, the impedance seen from the power AMP (Z_{in} in Fig. 4.7) is matched to 50 Ω, and 1 W power is transmitted from the power AMP using the resonant frequency of 13.56 MHz. The S-parameter simulation results are shown in Fig. 4.9. Under the initial conditions, the bold line of Fig. 4.9 indicates that the power is transferred well with S_{11} equal to -38 dB. The transfer distance is changed from 0.6 m to 0.2 m by moving the receiver coil. As a result, the impedance (Z_{in}) is increased to 785.1 Ω, and S_{21} is changed to -15 dB, indicating that power is not properly transferred at 13.56 MHz.

The power efficiency degradation can also be observed by measuring the current at the transmitter side. Figure 4.10 shows the current measured with the transfer distances of 0.2 and 0.6 m. There are large currents under the initial impedance matching conditions with a transfer distance of 0.6 m (bold line in Fig. 4.10), but when the distance is changed to 0.2 m, the measured current becomes small.

Figure 4.11 presents an equivalent circuit model of WPT transceiver shown in Fig. 4.7. On the transmitter side, the power amplifier and power source can be modeled as a current source parallel with an impedance using Norton's theorem. The matching network is also represented with a combination of series (C_{S1}) and

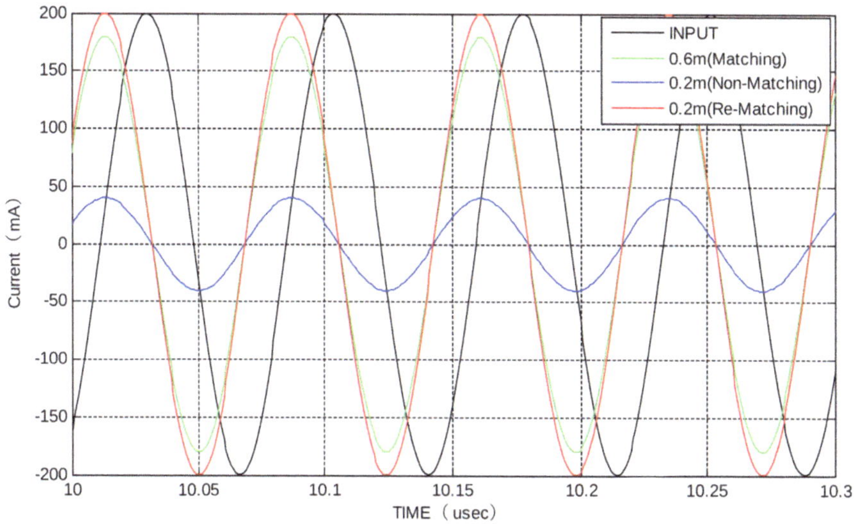

Fig. 4.10 Current waves measured at the transmitter with distances of 0.2 and 0.6 m

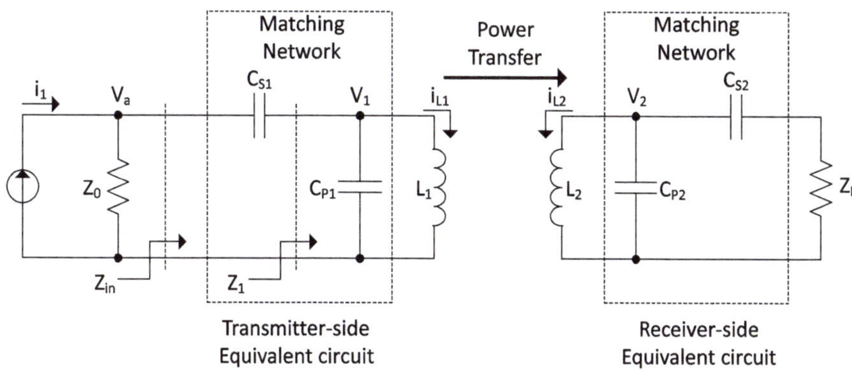

Fig. 4.11 Equivalent circuit model between transmitter and receiver

parallel (C_{P1}) capacitances. The impedance seen in the power amplifier (Z_{in}) should be perfectly matched with Z_0 (= 50 Ω) for maximum PTE. The matching network on the receiver side can also be modeled as a combination of parallel (C_{P2}) and series (C_{S2}) capacitances, and the rest of the receiver block is simply characterized as a load impedance (Z_L).

In the following, based on the circuit analysis, it is shown how the impedance (Z_{in}) is affected by the mutual inductance (M_{12}) changes mainly due to the varying distance between the transmitter and receiver. Later, it also show that the capacitances in the transmitter matching network (C_{S1} and C_{P1}) can be efficiently reconfigured to rematch the impedance, finally reaching the condition: $Z_0 = Z_{\text{in}} = 50\,\Omega$ under varying distance.

In Fig. 4.11, the impedance Z_{in} is represented as follows:

$$\frac{1}{Z_{in}} = \frac{i_1}{V_a} - \frac{1}{Z_0} = j\omega C_{S1}\left(1 - \frac{V_1}{V_a}\right) \tag{4.23}$$

The following two equations can be derived on the receiver side:

$$V_2 = j\omega L_2 i_{L2} + j\omega M i_{L1} \tag{4.24}$$

$$i_{L2} + j\omega C_{P2} V_2 + \frac{V_2}{\frac{1}{j\omega C_{S2}} + Z_L} = 0 \tag{4.25}$$

Let us refer the admittance Y_2 as follows:

$$Y_2 = j\omega C_{P2} + \frac{1}{\frac{1}{j\omega C_{S2}} + Z_L} \tag{4.26}$$

Then, (4.25) can be written as

$$i_{L2} + Y_2 V_2 = 0 \tag{4.27}$$

From the aforementioned equations, V_2 can be expressed with the current of the power receiver and the admittance Y_2 as

$$V_2 = -\frac{i_{L2}}{Y_2} \tag{4.28}$$

Now, (4.24) can be written as

$$j\omega L_2 i_{L2} + j\omega M i_{L1} + \frac{i_{L2}}{Y_2} = 0 \tag{4.29}$$

From (4.29), Z_2 is defined as

$$Z_2 = j\omega L_2 + \frac{1}{Y_2} \tag{4.30}$$

$$j\omega M i_{L1} = -Z_2 i_{L2} \tag{4.31}$$

Using (4.31), V_1 can be expressed as follows:

$$V_1 = \left[j\omega L_1 i_{L1} + j\omega M\left(\frac{-j\omega M}{Z_2}\right) i_{L1}\right] = \left[j\omega L_1 + \frac{\omega^2 M^2}{Z_2}\right] i_{L1} \tag{4.32}$$

Here, Z_M can be written as

$$Z_M = \frac{\omega^2 M^2}{Z_2} = \frac{\omega^2 k_{12}^2 L_1 L_2}{\frac{1-\omega^2 L_2 (C_{S2}+C_{P2})}{(1-\omega^2 L_2 C_{P2})(j\omega C_{S2})} + Z_L} = \frac{\omega^2 k_{12}^2 L_1 L_2}{Z_L} \tag{4.33}$$

For the above (4.33) on Z_M to be valid, the following two equations should be satisfied:

$$M = M_{12} = M_{21} = k_{12}\sqrt{L_1 L_2} \tag{4.34}$$

$$1 - \omega^2 L_2 (C_{S2} + C_{P2}) = 0 \tag{4.35}$$

where κ_{12} represents the coupling coefficient between two coils, and M is the mutual inductance. (4.35) also show the resonant condition for two coils. Accordingly, the impedance Z_{in} can be expressed as follows:

$$\begin{aligned} Z_{in} &= \frac{1}{j\omega C_{P1}} \,//\, (j\omega L_1 + Z_M) + \frac{1}{j\omega C_{S1}} \\ &= \frac{1}{j\omega C_{P1}} \,//\, \left(j\omega L_1 + \frac{\omega^2 k_{12}^2 L_1 L_2}{Z_L} \right) + \frac{1}{j\omega C_{S1}} \end{aligned} \tag{4.36}$$

From (4.36), Z_{in} can be expressed using Z_M, which is also a function of the mutual inductance, M_{12}. In other words, when the distance changes, the mutual inductance M_{12} is changed, which gives rise to changes in the impedance Z_M. The modified Z_M value results in the changes in Z_{in}, thus incurring impedance mismatch and serious PTE degradation. Figure 4.12 illustrates the equivalent circuit, where a new impedance component Z_M is added to the transmitter side meaning that the modifications of mutual inductance M_{12} can be modeled as the changes in Z_M in the transmitter.

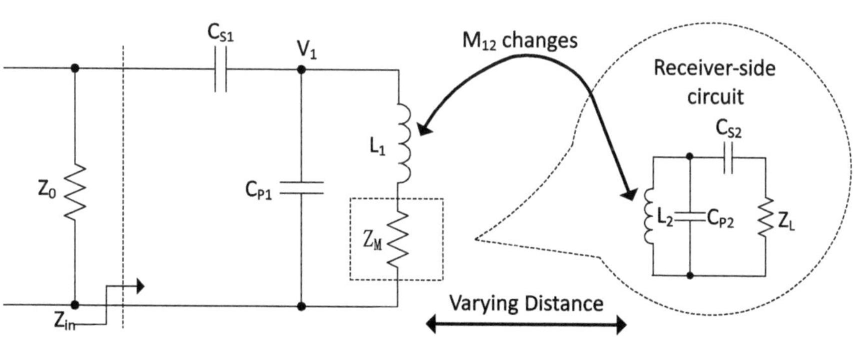

Fig. 4.12 Effect of changing distance on the power transmitter side, which can be modeled as the change of Z_M

In order to compensate for the changes in Z_M due to distance change, we can change the capacitance C_{P1} and C_{S1} such that the impedance (Z_{in}) is rematched to Z_0 (= 50Ω). When we change (4.36) into the real part of the impedance R and the imaginary part X, Z_{in} can be expressed as

$$\begin{aligned} Z_{in} = R + jX &= \frac{1}{j\omega C_{S1}} + \frac{1}{j\omega C_{P1}} \,//\, (j\omega L_1 + Z_M) \\ &= \frac{Z_M}{1 - 2\omega^2 C_{P1} L_1 + \omega^2 C_{P1}^2 Z_M^2 + \omega^4 C_{P1}^2 L_1^2} \\ &\quad + j\left(\frac{\omega L_1 - \omega^3 C_{P1} L_1^2 - \omega C_{P1} Z_M^2}{1 - 2\omega^2 C_{P1} L_1 + \omega^2 C_{P1}^2 Z_M^2 + \omega^4 C_{P1}^2 L_1^2} - \frac{1}{\omega C_{S1}}\right) \end{aligned} \tag{4.37}$$

In (4.37), for perfect impedance matching, the resistive term R should be equal to 50 Ω, and the reactance X should be zero. When the resistance term is adjusted to 50 Ω by modifying C_{P1}, the reactance X is affected, which may result in a resonant frequency shift. Since there is a C_{S1} term in the reactance term X, the modified reactance X due to the change in C_{P1} can be compensated by changing C_{S1}.

The previous part has shown that the impedance mismatch due to the distance change can be compensated by controlling the capacitance of the matching network in the transmitter. In this part, a capacitor matrix is presented, where the combination of series and parallel capacitances can be dynamically reconfigured to resolve the impedance mismatch. In the meanwhile, the resonant frequency is not changed when the capacitor matrix is used to track the impedance change.

The M × N capacitor matrix, where each capacitor can be connected either in parallel or series according to the switch array selection, is presented in Fig. 4.13a. For N capacitances with switches, there are 2^N different values of C_{P1} and C_{S1}. As an example, Fig. 4.13b shows a capacitor matrix with seven different capacitances. With the switch configuration presented in Fig. 4.13b, the equivalent circuit can be simplified into the one in Fig. 4.13c, where C_{S1} has the capacitance value of $C_3 + C_4 + C_5 + C_6$, and C_{P1} has a value of $C_1 + C_2 + C_7$. Using the capacitor matrix, the total impedance values can be dynamically changed by controlling the switches.

Based on the matrix with seven capacitors shown in Fig. 4.13b, where there are 128 combinations available, the impedance rematch process with varying distances has been tried based on the experimental setup specified in Table 4.1. Table 4.2 shows the optimal capacitance switch selections based on seven different capacitance values ($C_1, C_2, \ldots, C_7$) for seven different distances (0 m, 0.1 m, 0.2 m, . . . , 0.8 m). For each of the capacitance combinations with different distances, Z_{in} in Fig. 4.7 (the impedance seen from the power amplifier) is set close to 50Ω.

For example, for the impedance rematch with 0.2 m distance, the C_{S1} and C_{P1} values are supposed to be 30 and 97 pF, respectively. To realize the specific C_{S1} and C_{P1} values, C_3, C_4, C_5, and C_6 are connected in series, while C_1, C_2, and C_7 should be connected in parallel, as shown in Fig. 4.13b and c.

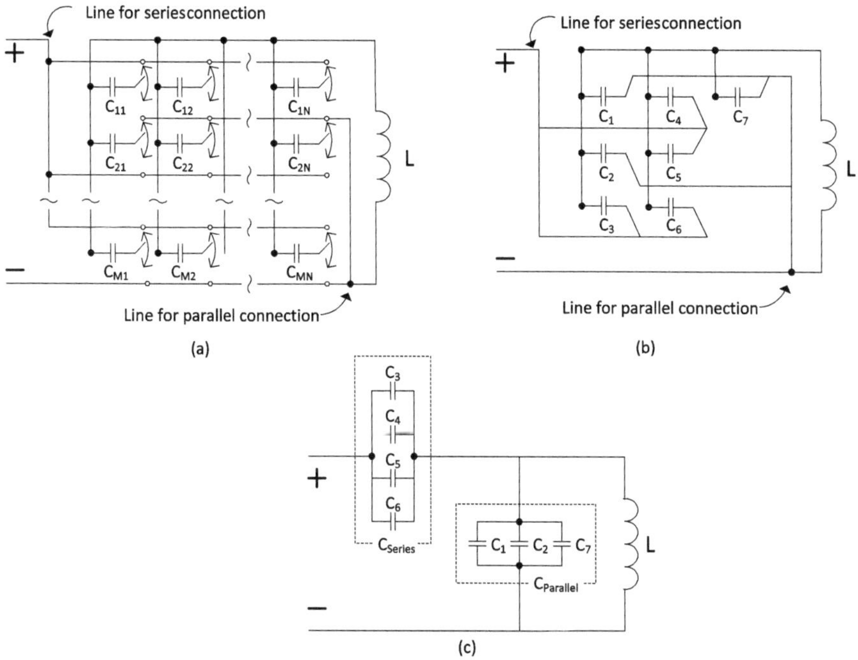

Fig. 4.13 (**a**) $M \times N$ capacitor matrix circuit. (**b**) Example of a capacitor matrix with seven capacitances. (**c**) Equivalent circuit of (b)

Table 4.2 Optimal capacitance switch configuration for various distances (0–0.8 m) when a capacitor matrix with seven capacitances is used

			Switch position						
			C_1	C_2	C_3	C_4	C_5	C_6	C_7
Distance (m)	C_{S1}	C_{P1}	64 pF	32 pF	16 pF	8 pF	4 pF	2 pF	1 pF
0	85	42	Series	Parallel	Series	Parallel	Series	Parallel	Series
0.1	61	66	Parallel	Series	Series	Series	Series	Parallel	Series
0.2	30	97	Parallel	Parallel	Series	Series	Series	Series	Parallel
0.3	15	112	Parallel	Parallel	Parallel	Series	Series	Series	Series
0.4	8	119	Parallel	Parallel	Parallel	Series	Parallel	Parallel	Parallel
0.5	5	122	Parallel	Parallel	Parallel	Parallel	Series	Parallel	Series
0.6	5	122	Parallel	Parallel	Parallel	Parallel	Series	Parallel	Series
0.7	4	123	Parallel	Parallel	Parallel	Parallel	Series	Parallel	Parallel
0.8	4	123	Parallel	Parallel	Parallel	Parallel	Series	Parallel	Parallel

When the capacitor-matrix is reconfigured to follow the impedance change, one of the key issues is that the resonant frequency should remain the same even though the impedance changes. In order to maintain the resonance, the reactance part, jX in (4.37), should be zero which can be expressed as

$$\frac{-Z_M^2\omega C_{P1} + \omega L_1 - \omega^3 L_1^2 C_{P1}}{1 - 2\omega^2 L_1 C_{P1} + \omega^4 L_1^2 C_{P1}^2 + \omega^2 C_{P1}^2 Z_M^2} = \frac{1}{\omega C_{S1}} \tag{4.38}$$

where C_{S1} represents the series capacitance, C_{P1} is the parallel capacitance, L_1 is the inductance, and Z_M represents the impedance induced by the mutual inductance shown in Fig. 4.11. To satisfy the condition shown in (4.38), the following equations should be satisfied:

$$\frac{1}{\omega^2} = L_1(C_{S1} + C_{P1}) \tag{4.39}$$

$$\frac{C_{P1}}{L_1} \cdot Z_M^2 \approx 0 \tag{4.40}$$

As shown in (4.39), in order to have the same resonant frequency before and after the distance change, the sum of the total capacitance in the matrix, $C_{S1} + C_{P1}$, should be kept the same, which means $\sum_{i=1}^{M}\sum_{j=1}^{N} C_{ij} = \frac{1}{\omega^2 L}$, where ω and L mean the resonant frequency and the inductance of the transmitter, respectively. As shown in Table 4.2, the C_{S1} + C_{P1} values are set to 127 pF to maintain the same resonant frequency. The physical meaning of (4.40) is that as the distance between the transmitter and receiver becomes smaller, since Z_M is getting larger, C_{P1} in the capacitor matrix should be negligibly small to satisfy (4.40). Here, since C_{P1} is very small, C_{S1} becomes larger (the switches configuration in the capacitor matrix is changed to increase C_{S1}) as C_{P1} + C_{S1} should be constant (127pF in our case). On the other hand, when the distance increases, Z_M and C_{S1} become smaller.

A series/parallel capacitor matrix is presented above, where the impedance can be dynamically reconfigured to track the optimum impedance matching point in case of transfer distance change. In the capacitor matrix, if there are a large number of capacitances, good resolution on the rematched impedance can be achieved.

4.2 Rectifiers

A conventional wireless energy transfer system essentially consists of two main platforms, transmitter and receiver stages [6]. High frequency ac energy is converted to dc in the receiver by rectifiers; an additional DC–DC converter stage may also be required depending on the application. The pickup energy at the receiver input terminals might be a high frequency current source or a high frequency voltage source related to the resonant compensation such as series, parallel, or LCC resonant tuning, etc. [7–9].

High frequency rectifiers must have fast switching diodes in order to perform better efficiency considering short forward and reverse recovery time [10, 11]. Due to temporary short circuit, a large reverse current might cause excessive power dissipation and ringing during reverse recovery time. Also, high voltage spikes might occur during the forward recovery time that the forward voltage drop can be higher than the normal on state forward voltage. Schottky diode is the most suitable high frequency rectifier component, providing the shortest reverse and forward recovery time. However, Schottky diode reverse break down voltage is relatively lower than the conventional low frequency diode rectifiers. Ultra-fast diodes could also be used if the proper Schottky diode was not available during the high frequency rectifier circuit design [12, 13].

This section presents high frequency rectifier topologies; Class D current driven rectifiers, Class D voltage driven rectifiers, and semi-bridgeless active rectifier topologies. Analyses of these rectifiers are given in the following subsections [14].

4.2.1 *Class D Current Driven Rectifiers*

The analytical treatments in Sects. 4.2.1 and 4.2.2 build on the authors' earlier treatment of Class D rectifiers in [15], but they are reorganized here for WPT receiver applications and adapted to the notation used in the figures of this section.

In many WPT receivers, the compensated pickup seen at the rectifier input is well approximated as a high-frequency sinusoidal current source. Under this condition, the rectifier belongs to the current-driven class. A particularly important subgroup is formed by circuits whose diode voltages are approximately square-wave while the diode currents are approximately half-sinusoidal, which is why they are commonly described as Class D current-driven rectifiers [15–18]. In this section, Class D current-driven bridge and synchronous rectifiers are presented considering the diode forward voltage V_f, forward resistance R_f, and filter-capacitor equivalent series resistance (ESR) r_C.

The analysis assumes a piecewise-linear diode model in the on state, negligible diode lead inductance and junction capacitance, zero minority-carrier lifetime for pn diodes, open-circuit diode off-state behavior with negligible leakage current, sinusoidal rectifier input current, and dc output ripple voltage V_r below 1%.

4.2.1.1 Bridge Rectifier

Figure 4.14 shows the Class D current-driven bridge rectifier with input current i_R and dc output current. The circuit consists of four diodes D_1, D_2, D_3, and D_4, a filter capacitor C_f, and dc load R_L. The receiver coupler system with resonant tuning is modeled as an ideal high-frequency sinusoidal current source i_R. When i_R is greater than zero, D_1 and D_4 conduct while D_2 and D_3 are off. When i_R is less than zero, the conduction pair reverses. The corresponding circuit voltage and current waveforms are given in Fig. 4.15. The voltage across the rectifier input terminals is approximately square-wave, with levels near $\pm V_O$. Each diode conducts for one half-cycle; the diode voltages are square-wave, and the diode currents are half-sinusoidal. The filter capacitor C_f maintains the output voltage V_O nearly

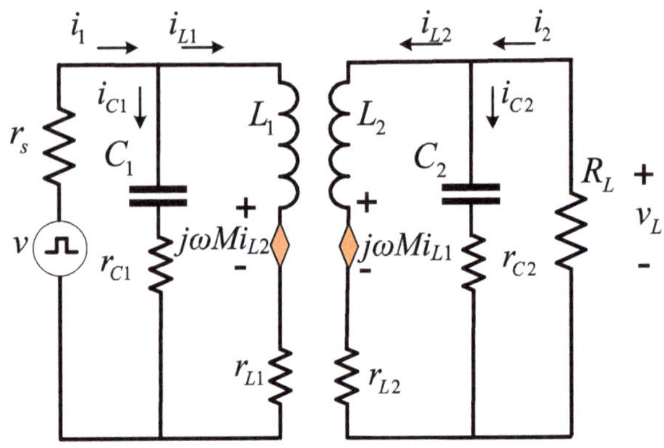

Fig. 4.14 A Class D current driven bridge rectifier circuit

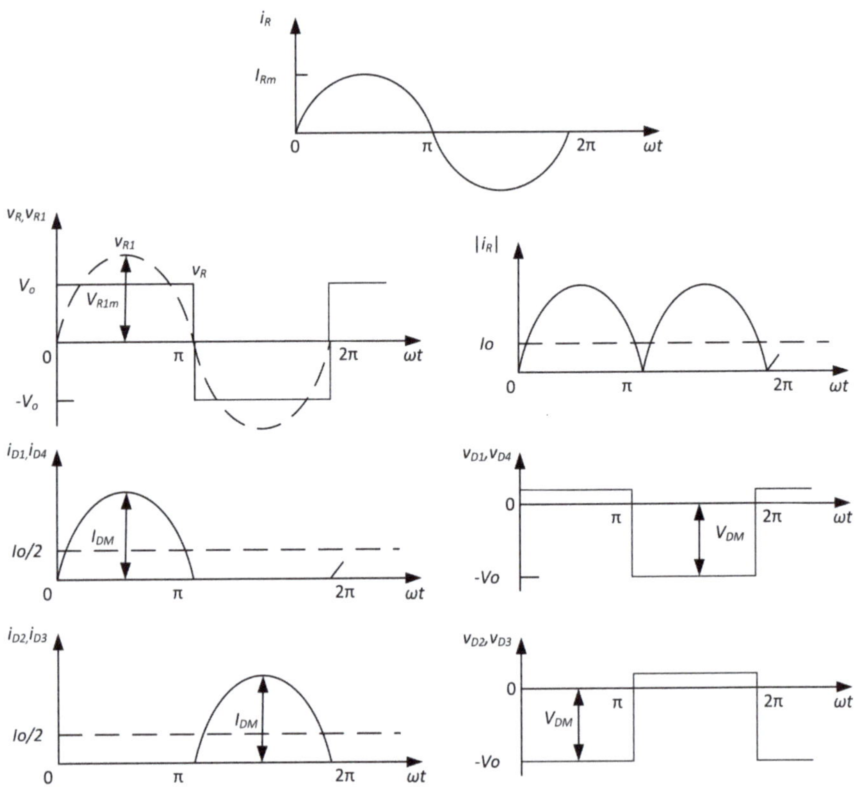

Fig. 4.15 Current and voltage waveforms in Class D current driven bridge rectifier circuit components

constant. As in the standard treatment of this rectifier class [15], the relatively low diode turn-off *di/dt* helps reduce reverse-recovery-related noise and loss, although current through the diode capacitances can still contribute to high-*dv/dt* effects. The governing relations are given below.

Assuming the input current i_R is purely sinusoidal, the half-sinusoidal current in the diodes $i_{D1,..,4}$ can be stated as,

$$i_{D1}, i_{D4} = \begin{cases} I_{Rm} \sin \omega t, for\ 0 < \omega t \leq \pi \\ 0, \quad for\ \pi < \omega t \leq 2\pi \end{cases} \tag{4.41}$$

$$i_{D2}, i_{D3} = \begin{cases} 0, \quad for\ 0 < \omega t \leq \pi \\ I_{Rm} \sin \omega t, \quad for\ \pi < \omega t \leq 2\pi \end{cases} \tag{4.42}$$

where I_{Rm} is the amplitude of i_R. Each diode current rms $I_{D,rms}$ and average values $I_{D,ave}$ are,

$$I_{D,rms} = \sqrt{\frac{1}{2\pi} \int_0^{2\pi} i_{D1,..,4}^2 d(\omega t)} = \frac{I_{Rm}}{2} \tag{4.43}$$

$$I_{D,ave} = \frac{1}{2\pi} \int_0^{2\pi} i_{D1,..,4}\, d(\omega t) = \frac{I_{Rm}}{\pi} \tag{4.44}$$

The output current I_O can be found as,

$$I_O = \frac{1}{2\pi} \int_0^{2\pi} (i_{D1}, i_{D4})\, d(\omega t) + \frac{1}{2\pi} \int_0^{2\pi} (i_{D2}, i_{D3})\, d(\omega t) = \frac{2I_{Rm}}{\pi} \tag{4.45}$$

The output current I_O is proportional to the amplitude of the input current I_{Rm}. The input high frequency ac to the output dc current transfer function M_{IR} can be calculated as,

$$M_{IR} = \frac{I_O}{I_{R,rms}} = \frac{\sqrt{2} I_O}{I_{Rm}} = \frac{2\sqrt{2}}{\pi} \tag{4.46}$$

where $I_{R,rms}$ is the rms value of the input current and equal to $\frac{I_{Rm}}{\sqrt{2}}$. The output voltage V_O can be written as,

$$V_O = I_O R_L = \frac{2 I_{Rm} R_L}{\pi} \tag{4.47}$$

The output voltage V_O can be regulated by adjusting the input current amplitude I_{Rm}. The input current i_R is in phase with the fundamental component v_{R1} of the input voltage v_R. The input power P_i can be specified as,

$$P_i = \frac{I_{Rm}^2 R_i}{2} = \frac{\pi^2 I_O^2 R_i}{8} = \frac{8V_O^2}{\pi^2 R_i} \tag{4.48}$$

where R_i is the rectifier input resistance. The output power P_O can be expressed as,

$$P_O = I_O^2 R_L = \frac{V_O^2}{R_L} \tag{4.49}$$

The square waveform of input voltage v_R can be defined neglecting the voltage drops across the diode as,

$$v_R = \begin{cases} V_O, & for\ 0 < \omega t \le \pi \\ -V_O, & for\ \pi < \omega t \le 2\pi \end{cases} \tag{4.50}$$

The rms value of the input voltage $V_{R,rms}$ can be seen as,

$$V_{R,rms} = \sqrt{\frac{1}{2\pi}\int_0^{2\pi} v_R^2 d(\omega t)} = V_O \tag{4.51}$$

The amplitude V_{R1m} and rms $V_{R1,rms}$ values of the input voltage fundamental component can be given as,

$$V_{R1m} = \frac{1}{\pi}\int_0^{2\pi} v_R \sin\omega t d(\omega t) = \frac{4V_O}{\pi} \tag{4.52}$$

$$V_{R1,rms} = \frac{V_{R1m}}{\sqrt{2}} = \frac{2\sqrt{2}V_O}{\pi} \tag{4.53}$$

Each of the diode peak forward current stresses I_{DM} can be described as,

$$I_{DM} = I_{Rm} = \frac{\pi I_O}{2} \tag{4.54}$$

and each of the diode peak reverse voltage stresses V_{DM} can be stated as,

$$V_{DM} = V_O \tag{4.55}$$

and the power output capability of the rectifier c_{pR} can be obtained as,

$$c_{pR} = \frac{P_O}{V_{DM} I_{DM}} = \frac{V_O I_O}{V_{DM} I_{DM}} = \frac{2}{\pi} \tag{4.56}$$

Using diode voltage and current peak values, the output power P_O can be achieved as,

$$P_O = c_{pR} V_{DM} I_{DM} \tag{4.57}$$

The efficiency of the rectifier η_R can be described by calculating the power losses of each component. The power dissipation in each of the diode P_D can be found using diode current rms $I_{D,rms}$ (4.43) and average $I_{D,ave}$ (4.44) values as,

$$P_D = V_f I_{D,ave} + I_{D,rms}^2 R_f = \frac{V_f I_O}{2} + \frac{\pi^2 I_O^2 R_f}{16} \tag{4.58}$$

The filter current i_{Cf} can be estimated assuming $\frac{1}{\omega C_f} \ll R_L$ as,

$$i_{Cf} = |i_R| - I_O \tag{4.59}$$

$$= I_{Rm} |\sin \omega t| - I_O = I_O \left(\frac{\pi}{2} |\sin \omega t| - 1 \right) \tag{4.60}$$

and rms value of the filter current $i_{Cf,rms}$ can be found as,

$$I_{Cf,rms} = \sqrt{\frac{1}{2\pi} \int_0^{2\pi} i_{Cf}^2 d(\omega t)} = I_O \sqrt{\frac{\pi^2}{8} - 1} \tag{4.61}$$

The power dissipation in the filter capacitor P_{rC} due to ESR is,

$$P_{rC} = I_{Cf,rms}^2 r_C = I_O^2 r_C \left(\frac{\pi^2}{8} - 1 \right) \tag{4.62}$$

The total power dissipation P_T from the input to the output of the rectifier can be found as,

$$P_T = 4P_D + P_{rC} = 2V_f I_O + \frac{\pi^2 I_O^2 R_f}{4} + I_O^2 r_C \left(\frac{\pi^2}{8} - 1 \right) \tag{4.63}$$

The overall system efficiency of the rectifier η_R circuit can be described as,

$$\eta_R = \frac{1}{1 + \frac{P_T}{P_O}} = \frac{1}{1 + \frac{2V_f}{V_O} + \frac{\pi^2 R_f}{4R_L} + \frac{r_C}{R_L} \left(\frac{\pi^2}{8} - 1 \right)} \tag{4.64}$$

From the rectifier efficiency point, the input resistance R_i across the rectifier input terminals can be computed as,

$$R_i = \frac{8R_L}{\pi^2 \eta_R} \tag{4.65}$$

The system voltage transfer function M_{VR} can be expressed as,

$$M_{VR} = \frac{V_O}{V_{R,rms}} = M_{IR}\frac{R_L}{R_i} = \frac{\pi \eta_R}{2\sqrt{2}} \tag{4.66}$$

Example 4.1 For the Class D current source bridge rectifier as given in Fig. 4.14, calculate the efficiency of the rectifier η_R, the voltage transfer function M_{VR}, and the input resistance R_i at V_O = 400 V and I_O = 2.5 A. The rectifier employs Schottky diodes with V_f = 0.6 V, R_f = 20 mΩ, and a filter capacitor with r_C = 10 mΩ. Also, find the voltage and current stresses across the diodes (Fig. 4.15).

Solution

The load resistance and output power of the rectifier can be found as,

$$R_L = \frac{V_O}{I_O} = \frac{400}{2.5} = 160\ \Omega \tag{4.67}$$

$$P_O = V_O I_O = 400 \times 2.5 = 1\ \text{kW} \tag{4.68}$$

The diode conduction losses can be calculated using (4.58) as,

$$P_D = \frac{V_f I_O}{2} + \frac{\pi^2 I_O^2 R_f}{16} = \frac{0.6 \times 2.5}{2} + \frac{\pi^2 \times 2.5^2 \times 20 \times 10^{-3}}{16} = 0.827\ \text{W} \tag{4.69}$$

The filter capacitor power loss can be obtained from (4.62) as,

$$P_{rC} = I_O^2 r_C \left(\frac{\pi^2}{8} - 1\right) = 2.5^2 \times 10^{-3} \times \left(\frac{\pi^2}{8} - 1\right) = 14.6\ \text{mW} \tag{4.70}$$

Using (4.63), the total losses are,

$$P_T = 4P_D + P_{rC} = 4 \times 0.827 + 14.6 \times 10^{-3} = 3.32\ \text{W} \tag{4.71}$$

The efficiency of the rectifier is from (4.64) as,

$$\eta_R = \frac{1}{1 + \frac{P_T}{P_O}} = \frac{1}{1 + \frac{3.31}{1 \times 10^3}} \times 100 = 99.7\% \tag{4.72}$$

The system voltage transfer function can be calculated by using (4.66) as,

$$M_{VR} = \frac{\pi \eta_R}{2\sqrt{2}} = \frac{\pi \times 0.997}{2\sqrt{2}} = 1.107 \tag{4.73}$$

From (4.65), the bridge rectifier input resistance can be computed as

$$R_i = \frac{8R_L}{\pi^2 \eta_R} = \frac{8 \times 160}{\pi^2 \times 0.997} = 130\ \Omega \tag{4.74}$$

Using (4.55) and (4.54), the diode voltage and current stresses can be found as,

$$V_{DM} = V_O = 400\ \text{V} \tag{4.75}$$

$$I_{DM} = \frac{\pi I_O}{2} = \frac{\pi \times 2.5}{2} = 3.93\ \text{A} \tag{4.76}$$

The bridge rectifiers are suitable for high output voltage applications. Two diodes conduct in each half-cycle and the forward voltage drop can cause the efficiency degradation at the low output voltage and high current. Since the voltage and current stresses are proper for the selected high frequency diode, the bridge rectifier is a strong candidate for wireless energy transfer systems.

4.2.1.2 Synchronous Rectifier

At low dc output voltage and high output current conditions, high frequency diode rectifiers have efficiency disadvantages due to diode forward voltage V_f, causing high conduction losses at high current. This condition increases temperature of the rectifier system and it might require large cooling system with increasing the size of the rectifier and unreliability. Therefore, synchronous rectifiers are proposed in several applications to increase the system efficiency by replacing diodes with low r_{DSon} resistance power FETs [19–21].

A non-controlled Class D current driven synchronous rectifier is described for wireless power transfer systems in Fig. 4.16. The synchronous rectifier system comprises a center-tapped coupler coil with compensation circuit, MOSFETs, filter capacitor, and load. In order to provide rectification, the MOSFETs are conducted in synchronism according to the ac current polarity. The current rectifying transition can be stated for each of MOSFET during positive and negative cycles that S_1 is on state and S_2 is off when the rectifier input current i_R is greater than zero, while S_1 is off and S_2 is on state for i_R lower than zero. The switch current flows through the MOSFET channel if the MOSFET drain to source voltage v_{DS} is less than the forward voltage V_f of anti-parallel diode of MOSFETs. Due to low r_{DSon} resistance of MOSFETs, a low power dissipation might be achieved depending on the switching losses at voltage and current amplitudes. Moreover, the MOSFETs

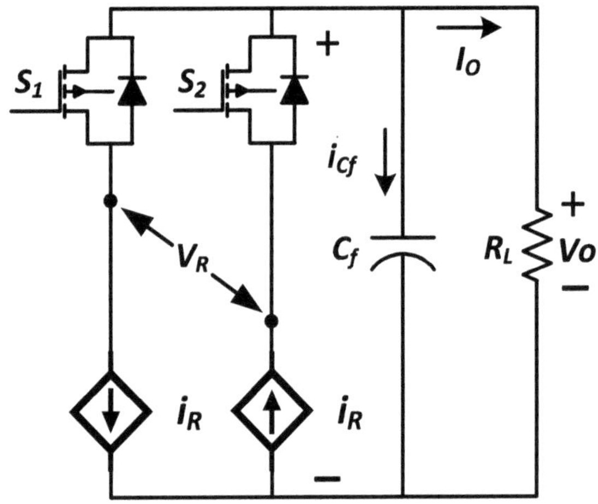

Fig. 4.16 A non-controlled Class D current driven synchronous rectifier

leakage current is much lower than the rectifier diodes. However, a driver circuit is required to turn-on and turn-off the MOSFETs. An additional auxiliary power supply is required, which consumes energy and increases the circuit complexity. The energy losses of auxiliary power supply might lead to the total conduction and switching losses of MOSFETs, especially for the high frequency transitions.

As assumed the rectifier input current i_R is sinusoidal, S_1 current i_{S1} is half sine wave for $0 < \omega t \leq \pi$ and zero for $\pi < \omega t \leq 2\pi$, S_2 current i_{S2} is zero for $0 < \omega t \leq \pi$ and half sine wave for $\pi < \omega t \leq 2\pi$. The output current I_O can be calculated as,

$$I_O = \frac{1}{2\pi}\int_0^{2\pi} i_{S1} d(\omega t) + \frac{1}{2\pi}\int_0^{2\pi} i_{S2} d(\omega t) = \frac{2I_{Rm}}{\pi} \tag{4.77}$$

where I_{Rm} is the amplitude of i_R. The current gain M_{IR}, input and output voltage and power equations are the same as for the bridge rectifier, from (4.46) to (4.79). The MOSFET current stresses I_{DM}, voltage stresses V_{DM}, and power output capability c_{pR} can be stated as,

$$I_{DM} = I_{Rm} = \frac{\pi I_O}{2} \tag{4.78}$$

$$V_{DM} = 2V_O \tag{4.79}$$

$$c_{pR} = \frac{P_O}{V_{DM} I_{DM}} = \frac{V_O I_O}{V_{DM} I_{DM}} = \frac{1}{\pi} \tag{4.80}$$

The gate drive power losses can be found by using the gate charge Q_g information from the MOSFETs datasheet. The gate power losses P_g to charge and discharge each of the MOSFET can be computed as,

$$P_g = Q_g V_{\text{GSpp}} f_s \tag{4.81}$$

where V_{GSpp} is the peak-to-peak gate voltage and f_s is the MOSFET switching frequency. The rectifier synchronous MOSFETs is normally operated in ZVS condition and assumed that there is no turn-on switching losses. However, the turn-off losses $P_{t\text{-}off}$ of each MOSFET can be approximated as,

$$P_{t-off} = 2V_O \left|i_{S1,2} \sin\theta\right| f_s \left(\frac{2t_r}{3} + t_f\right) \tag{4.82}$$

where θ is the phase difference between switch voltage and current signals, t_r is the turning-off voltage rise time and t_f is the current fall time. The gate and turn-off power losses are proportional to the frequency and might be equal to the conduction losses of MOSFET at high frequencies. Thus, synchronous rectifier is not suggested at very high frequency and high output voltage applications. The conduction losses of each MOSFET can be derived by rms value of the current $I_{S,rms}$ as,

$$I_{S,rms} = \sqrt{\frac{1}{2\pi}\int_0^{2\pi} i_{S1,2}^2 d(\omega t)} = \frac{\pi I_O}{4} \tag{4.83}$$

$$P_{DS} = I_{S,rms}^2 r_{\text{DSon}} = \frac{\pi^2 I_O^2 r_{\text{DSon}}}{16} \tag{4.84}$$

The total losses P_T of the rectifier can be found with the filter capacitor losses P_{rC} as given in (4.62), and are,

$$\begin{aligned} P_T &= 2P_g + 2P_{t-off} + 2P_{DS} + P_{rC} \\ &= 2Q_g V_{\text{GSpp}} f_s + 4V_O \left|i_{S1,2} \sin\theta\right| f_s \left(\frac{2t_r}{3} + t_f\right) \\ &+ \frac{\pi^2 I_O^2 r_{\text{DSon}}}{8} + I_O^2 r_C \left(\frac{\pi^2}{8} - 1\right) \end{aligned} \tag{4.85}$$

The rectifier efficiency η_R can be found as,

$$\begin{aligned} \eta_R &= \frac{1}{1 + \frac{P_T}{P_O}} \\ &= \frac{1}{1 + \frac{2Q_g V_{\text{GSpp}} f_s}{P_O} + \frac{4V_O \left|i_{S1,2}\sin\theta\right| f_s \left(\frac{2t_r}{3} + t_f\right)}{P_O} + \frac{\pi^2 I_O^2 r_{\text{DSon}}}{8} + \frac{r_C}{R_L}\left(\frac{\pi^2}{8} - 1\right)} \end{aligned} \tag{4.86}$$

The input resistance R_i of the system can be given as,

$$R_i = \frac{8R_L}{\pi^2 \eta_R} \quad (4.87)$$

The rectifier voltage transfer function M_V is

$$M_{VR} = \frac{V_O}{V_{R,rms}} = M_{IR}\frac{R_L}{R_i} = \frac{\pi \eta_R}{2\sqrt{2}} \quad (4.88)$$

For WPT applications, the current-driven bridge rectifier remains attractive because of its structural simplicity, modest device count, and straightforward implementation with a single output capacitor. Its main drawback appears in low-output-voltage, high-output-current receivers, where the combined effect of diode forward voltage V_f, diode forward resistance R_f, and capacitor ESR r_C can noticeably reduce rectifier efficiency η_R. In that operating region, the center- tapped synchronous alternative can improve efficiency because its low r_{DSon} values reduce conduction loss.

This benefit is obtained at the expense of higher implementation complexity, a center-tapped receiver winding, and higher MOSFET voltage stress V_{DM} than in the bridge rectifier. In addition, at high operating frequency f_s and high output current I_O, gate-drive, switching, and conduction losses can offset the expected synchronous- rectifier advantage. Consequently, the bridge rectifier is usually the more practical choice for higher-output-voltage WPT receivers, whereas the center-tapped synchronous rectifier is most attractive for low-voltage, high-current loads [15].

4.2.2 *Class D Voltage-Driven Rectifiers*

The voltage-driven case arises when the receiver compensation network presents the rectifier with an approximately sinusoidal voltage source. In this case, the receiver operates as a voltage-driven rectifier [22, 23]. When the diode voltages are half-sinusoidal and the diode currents are approximately square-wave, the circuit belongs to the Class D voltage-driven family, consistent with the classical treatment in [15]. A bridge rectifier is the most common topology in this group, and in this section bridge and synchronous realizations are examined for WPT receivers.

The analysis assumes a piecewise-linear diode on-state model with forward voltage V_f and forward resistance R_f, as well as open-circuit diode off-state behavior. Diode junction capacitance and lead inductance are considered negligible, and the minority-carrier lifetime for pn diodes is assumed to be zero. The rectifier input terminals are excited with a sinusoidal voltage, with V_f smaller than the rectifier input-voltage amplitude V_{Rm}. An output-filter inductance L_f is assumed large enough to keep the ac current ripple small.

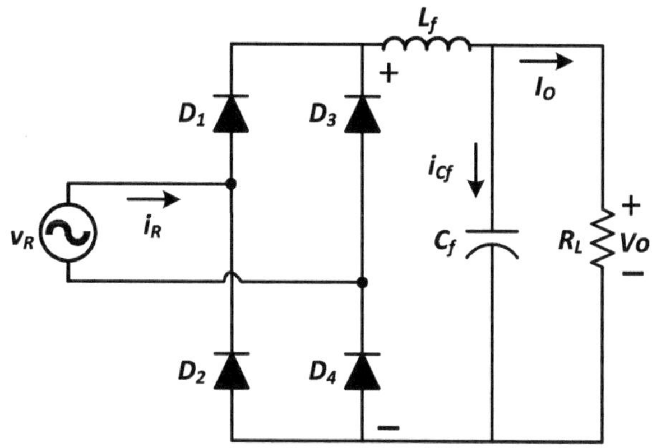

Fig. 4.17 A Class D voltage driven bridge rectifier circuit

4.2.2.1 Class D Bridge Rectifier

Figure 4.17 shows the Class D voltage-driven bridge rectifier excited by a high-frequency sinusoidal source. During the positive half-cycle, D_1 and D_4 conduct while D_2 and D_3 are off. During the negative half-cycle, D_2 and D_3 conduct while D_1 and D_4 are off. The corresponding circuit voltage and current waveforms are shown in Fig. 4.18. Under this operating pattern, the diode voltages are half-sinusoidal, the diode currents are square-wave, and each device conducts for one half-cycle, as in the standard Class D voltage-driven bridge rectifier description [15]. The governing relations are given below.

As assumed the input voltage v_R is pure sinusoidal, the diode voltages $v_{D1,..,4}$ can be described as,

$$v_{D1}, v_{D4} = \begin{cases} -V_{Rm} \sin \omega t, for\ 0 < \omega t \leq \pi \\ 0, \quad for\ \pi < \omega t \leq 2\pi \end{cases} \tag{4.89}$$

$$v_{D2}, v_{D3} = \begin{cases} 0, \quad for\ 0 < \omega t \leq \pi \\ -V_{Rm} \sin \omega t, \quad for\ \pi < \omega t \leq 2\pi \end{cases} \tag{4.90}$$

where V_{Rm} is the amplitude of the input voltage v_R. The output voltage V_O can be computed as,

$$V_O = \frac{1}{2\pi} \int_0^{2\pi} (v_{D1}, v_{D3})\, d(\omega t) + \frac{1}{2\pi} \int_0^{2\pi} (v_{D2}, v_{D4})\, d(\omega t) = \frac{2V_{Rm}}{\pi} \tag{4.91}$$

The output voltage V_O is proportional to the amplitude of the input voltage V_{Rm} and thus, V_O can be regulated by V_{Rm}. Each diode current rms $I_{D,rms}$ and average values $I_{D,ave}$ can be found as,

$$I_{D,rms} = \sqrt{\frac{1}{2\pi} \int_0^{2\pi} i_{D1,\ldots,4}^2 d(\omega t)} = \frac{I_O}{\sqrt{2}} \tag{4.92}$$

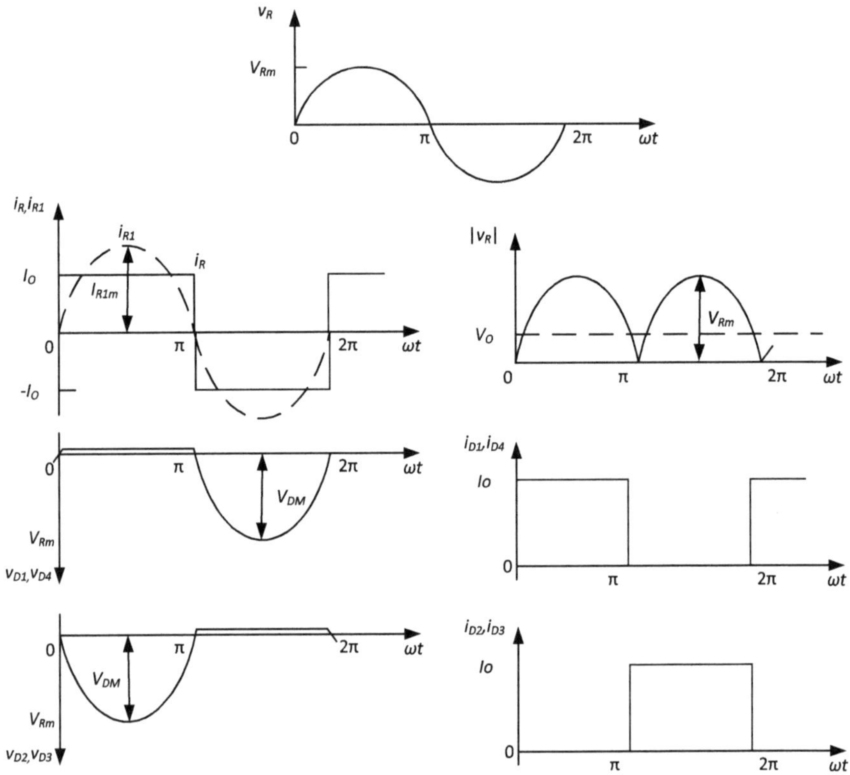

Fig. 4.18 Current and voltage waveforms in Class D voltage driven bridge rectifier circuit components

$$I_{D,ave} = \frac{1}{2\pi} \int_0^{2\pi} i_{D1,..,4}\, d(\omega t) = \frac{I_O}{2} \tag{4.93}$$

The square waveform of input current i_R can be stated as,

$$i_R = \begin{cases} I_O, & for\ 0 < \omega t \leq \pi \\ -I_O, & for\ \pi < \omega t \leq 2\pi \end{cases} \tag{4.94}$$

The rms value of the input current $I_{R,rms}$ can be extracted as,

$$I_{R,rms} = \sqrt{\frac{1}{2\pi} \int_0^{2\pi} i_R^2 d(\omega t)} = I_O \tag{4.95}$$

The amplitude I_{R1m} and rms $I_{R1,rms}$ values of the input current fundamental component can be obtained as,

$$I_{R1m} = \frac{1}{\pi} \int_0^{2\pi} i_R \sin \omega t d(\omega t) = \frac{4I_O}{\pi} \tag{4.96}$$

$$I_{R1,rms} = \frac{I_{R1m}}{\sqrt{2}} = \frac{2\sqrt{2}I_O}{\pi} \tag{4.97}$$

The output dc current to the input high frequency ac transfer function M_{IR} can be approximated as,

$$M_{IR} = \frac{I_O}{I_{R1,rms}} = \frac{\pi}{2\sqrt{2}} \tag{4.98}$$

The output current I_O can be specified as,

$$I_O = \frac{V_O}{R_L} = \frac{2V_{Rm}}{\pi R_L} \tag{4.99}$$

The input voltage v_R is in phase with the fundamental component i_{R1} of the input current i_R. The input power P_i can be written as,

$$P_i = \frac{V_{Rm}^2}{2R_i} = \frac{V_O^2 \pi^2}{8R_i} = \frac{8I_O^2 R_i}{\pi^2} \tag{4.100}$$

where R_i is the rectifier input resistance. The output power P_O can be described as,

$$P_O = \frac{V_O^2}{R_L} = I_O^2 R_L \tag{4.101}$$

Each of the diode peak forward voltage stresses V_{DM} can be defined as,

$$V_{DM} = V_{Rm} = \frac{V_O \pi}{2} \tag{4.102}$$

and each of the diode peak reverse current stresses I_{DM} can be found as,

$$I_{DM} = I_O \tag{4.103}$$

and the power output capability of the rectifier c_{pR} can be extracted as,

$$c_{pR} = \frac{P_O}{V_{DM} I_{DM}} = \frac{V_O I_O}{V_{DM} I_{DM}} = \frac{2}{\pi} \tag{4.104}$$

Using diode voltage and current peak values, the output power P_O can be stated as,

$$P_O = c_{pR} V_{DM} I_{DM} \tag{4.105}$$

The efficiency of the rectifier η_R can be described by calculating the power losses of each component. The power dissipation in each of the diode P_D and output filter inductor P_{rL} can be found using diode current rms $I_{D,rms}$ (4.92), average $I_{D,ave}$ (4.93), and output current I_O values as,

$$P_D = V_f I_{D,ave} + I_{D,rms}^2 R_f = \frac{V_f I_O}{2} + \frac{I_O^2 R_f}{2} \tag{4.106}$$

$$P_{rL} = I_O^2 r_L \tag{4.107}$$

where r_L is the dc ESR value of the filter inductor L_f. The filter current i_{Cf} can be estimated ignoring the ripple current and flows through the filter capacitor C_f. The voltage across the filter inductor v_{Lf} is,

$$\begin{aligned} v_{Lf} &= |v_R| - V_O \\ &= V_{Rm} |\sin \omega t| - V_O = V_O \left(\frac{\pi}{2} |\sin \omega t| - 1 \right) \end{aligned} \tag{4.108}$$

The inductor ac and dc currents i_{Lf} can be given as,

$$i_{Lf} = \frac{1}{\omega L_f} \int_0^{\omega t} v_L d(\omega t) + I_O = \frac{V_O}{\omega L_f} \left(-\frac{\pi}{2} \cos \omega t - \omega t + \frac{\pi}{2} \right) + I_O \tag{4.109}$$

and filter capacitor ac current i_{Cf} and rms values $i_{Cf,rms}$ can be calculated as,

$$i_{Cf} = i_{Lf} - I_O \tag{4.110}$$

$$I_{Cf,rms} = \sqrt{\frac{1}{2\pi} \int_0^{2\pi} i_{Cf}^2 d(\omega t)} = \frac{I_O R_L}{f_s L_f} \sqrt{\frac{5}{24} - \frac{2}{\pi^2}} \tag{4.111}$$

The ac power dissipation in the output filter P_{rLC} due to ac inductor and capacitor ESR r_{LC} values as,

$$P_{rLC} = I_{Cf,rms}^2 r_{LC} = \left(\frac{5}{24} - \frac{2}{\pi^2} \right) \frac{{I_O}^2 {R_L}^2}{{f_s}^2 {L_f}^2} r_{LC} \tag{4.112}$$

The total power dissipation P_T from the input to the output of the rectifier can be found as,

$$\begin{aligned} P_T &= 4P_D + P_{rL} + P_{rLC} \\ &= 2V_f I_O + 2I_O^2 R_f + I_O^2 r_L + \left(\frac{5}{24} - \frac{2}{\pi^2} \right) \frac{{I_O}^2 {R_L}^2 r_{LC}}{{f_s}^2 {L_f}^2} \end{aligned} \tag{4.113}$$

The overall system efficiency of the rectifier η_R circuit can be stated as,

$$\eta_R = \frac{1}{1+\frac{P_T}{P_O}} = \frac{1}{1+\frac{2V_f}{V_O}+\frac{2R_f}{R_L}+\frac{r_L}{R_L}+\left(\frac{5}{24}-\frac{2}{\pi^2}\right)\frac{R_L r_{LC}}{{f_s}^2 {L_f}^2}} \tag{4.114}$$

The input resistance R_i across the rectifier input terminals can be found as,

$$R_i = \frac{\pi^2 R_L}{8\eta_R} \tag{4.115}$$

The rectifier voltage transfer function M_{VR} can be described as,

$$M_{VR} = \frac{V_O}{V_{R,rms}} = M_{IR}\frac{R_L}{R_i} = \frac{2\sqrt{2}}{\pi\eta_R} \tag{4.116}$$

Example 4.2 For the Class D voltage source bridge rectifier as given in Fig. 4.17, find the efficiency of the rectifier η_R, the voltage transfer function M_{VR}, and the input resistance R_i at V_O = 24 V and I_O = 3 A. The rectifier operating frequency f_s is 150 kHz and it employs Schottky diodes with V_f = 0.5 V, R_f = 35 mΩ. The output filter inductor L_f value is 20 μH with dc ESR r_L 15 mΩ and ac ESR r_l 0.75 Ω, and the output filter capacitor ESR r_C value is 10 mΩ.

Solution

The load resistance of the rectifier is $R_L = V_O/I_O$ = 8 Ω and the output power is $P_O = V_O x I_O$ = 72 W. Each diode conduction loss P_D is found using (4.106) as

$$P_D = \frac{0.5\times 3}{2}+\frac{3^2\times 0.035}{2} = 0.908\ \text{W} \tag{4.117}$$

The filter inductor dc ESR loss P_{rL} is,

$$P_{rL} = 3^2\times 0.015 = 0.135\ \text{W} \tag{4.118}$$

The output filter ac power losses P_{rLC} is,

$$P_{rLC} = \left(\frac{5}{24}-\frac{2}{\pi^2}\right)\times\frac{3^2\times 8^2}{\left(150\times 10^3\right)^2\times\left(20\times 10^{-6}\right)^2}\times(0.75+0.01)$$
$$= 0.277\ \text{W} \tag{4.119}$$

The bridge rectifier efficiency η_R is calculated from (4.113) and (4.114) as,

$$\eta_R = \frac{1}{1+\frac{P_T}{P_O}} = \frac{1}{1+\frac{4\times 0.908+0.135+0.277}{72}}\times 100 = 94.7\% \tag{4.120}$$

Using (4.116), the dc output to ac input voltage transfer function of the rectifier M_{VR} is,

$$M_{VR} = \frac{2\sqrt{2}}{\pi \times 0.947} = 0.95 \tag{4.121}$$

Using (4.45), the input resistance of the rectifier R_i is,

$$R_i = \frac{\pi^2 \times 8}{8 \times 0.947} = 10.422\ \Omega \tag{4.122}$$

This example shows that the bridge rectifier efficiency can be degraded with the low output applications. The main losses are observed in the diode conduction losses due to two diodes conducting in each of half cycle and the forward voltage drop. The conduction losses can be reduced with synchronous rectifiers for low output voltage applications.

4.2.2.2 Synchronous Rectifier

Synchronous rectifiers are proposed to alleviate the rectifier efficiency that diode forward voltage might cause to increase the conduction losses of the rectifier at low output voltage and high output current systems [24]. An uncontrolled Class D voltage driven center taped rectifier system is illustrated in Fig. 4.19. The rectifier diodes are replaced with power MOSFET that can be turned on or off based on the input voltage polarity in positive and negative sections.

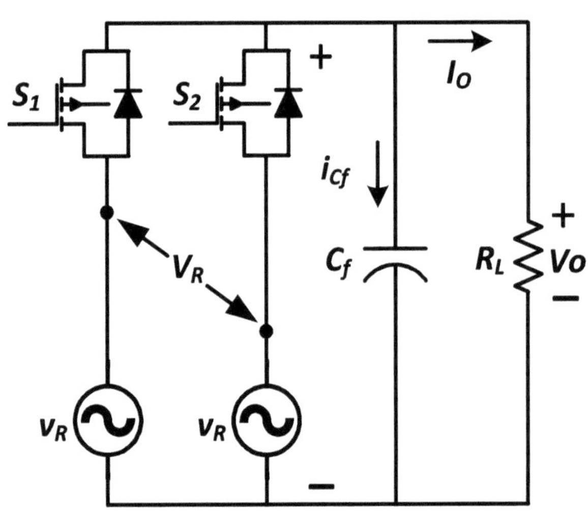

Fig. 4.19 An uncontrolled Class D voltage driven synchronous rectifier

The rectifier output voltage average value can be calculated as,

$$V_O = \frac{1}{2\pi}\int_0^{2\pi} v_R d(\omega t) = \frac{1}{2\pi}\int_0^{2\pi} V_{Rm}\left|\sin\omega t\right| d(\omega t) = \frac{2V_{Rm}}{\pi} \tag{4.123}$$

where V_{Rm} is the amplitude of v_R. The output voltage V_O is proportional to the input voltagc amplitude V_{Rm}. The input current i_R, current gain M_{IR}, output current, and power equations are the same as bridge rectifier from (4.96) to (4.104). The MOSFET voltage stresses V_{DM}, current stresses I_{DM}, and power output capability c_{pR} can be stated as,

$$V_{DM} = 2V_{Rm} = \pi V_O \tag{4.124}$$

$$I_{DM} = I_O \tag{4.125}$$

$$c_{pR} = \frac{P_O}{V_{DM}I_{DM}} = \frac{V_O I_O}{V_{DM}I_{DM}} = \frac{1}{\pi} \tag{4.126}$$

By computing each component power losses of the rectifier, the efficiency of the system η_R can be extracted. The power losses of gate drive P_g is the same as described in (4.81). MOSFET turning-off losses $P_{t\text{-}off}$ can be approximated for each component as,

$$P_{t-off} = 2I_O\left|v_{S1,2}\sin\theta\right| f_s\left(\frac{2t_r}{3} + t_f\right) \tag{4.127}$$

where θ is the phase difference between switch voltage and current signals, t_r is the turning-off voltage rise time and t_f is the current fall time. The conduction losses of each MOSFET can be derived by rms value of the current $I_{S,rms}$ as,

$$P_{DS} = I_{S,rms}^2 r_{\text{DSon}} = \frac{I_O^2 r_{\text{DSon}}}{2} \tag{4.128}$$

The output filter losses can be found as in the bridge rectifier, as given in (4.111) through (4.122). The total rectifier losses P_T can be obtained as,

$$\begin{aligned} P_T &= 2P_g + 2P_{t-off} + 2P_{DS} + P_{rL} + P_{rLC} \\ &= 2Q_g V_{\text{GSpp}} f_s + 4I_O\left|v_{S1,2}\sin\theta\right| f_s\left(\frac{2t_r}{3} + t_f\right) \\ &\quad + I_O^2 r_{\text{DSon}} + I_O^2 r_L + \left(\frac{5}{24} - \frac{2}{\pi^2}\right)\frac{{I_O}^2{R_L}^2 r_{LC}}{{f_s}^2{L_f}^2} \end{aligned} \tag{4.129}$$

The rectifier efficiency η_R is,

$$\eta_R = \frac{1}{1+\frac{P_T}{P_O}} = \frac{1}{1+\frac{2Q_g V_{GSpp} f_s}{P_O}+\frac{4I_O|v_{S1,2}\sin\theta| f_s\left(\frac{2t_r}{3}+t_f\right)}{P_O}+\frac{r_{DSon}}{R_L}+\frac{r_L}{R_L}+\left(\frac{5}{24}-\frac{2}{\pi^2}\right)\frac{R_L r_{LC}}{f_s{}^2 L_f{}^2}} \tag{4.130}$$

The rectifier input resistance R_i can be written as,

$$R_i = \frac{\pi^2 R_L}{8\eta_R} \tag{4.131}$$

The voltage transfer function M_V of the rectifier is

$$M_{VR} = \frac{V_O}{V_{R,rms}} = M_{IR}\frac{R_L}{R_i} = \frac{2\sqrt{2}}{\pi\eta_R} \tag{4.132}$$

In contrast to the current-driven case, the Class D voltage-driven rectifier uses a second-order output filter composed of L_f and C_f, and the dc output voltage V_O follows the amplitude of the applied ac input voltage V_{Rm}. In both the bridge and synchronous realizations, each device conducts for one half-cycle under the assumed operating conditions. The output-filter corner frequency f_c $(=\frac{1}{2\pi\sqrt{L_f C_f}})$ is independent of the output load R_L. A practical concern in diode implementations is the higher peak reverse-recovery current associated with large turn-off *di/dt*, which can increase both noise and loss.

For this reason, synchronous rectification is especially attractive in low-output-voltage applications, where low R_{DSon} can reduce conduction loss. At the same time, gate-driver and switching losses P_g and $P_{t\text{-}off}$ may become dominant depending on the operating voltage and current. Thus, the choice between diode and synchronous implementations is set primarily by output-voltage level, output-current level, and operating frequency [15].

4.2.3 Semi-bridgeless Rectifiers

In wireless power transfer applications, impedance matching is crucial to improve energy transfer efficiency and to keep the system at the optimal load with soft switching, especially for misalignment tolerance and variable load applications. Although reactive power compensation helps to improve system performance during coil misalignment, maintaining a wide load range while achieving high efficiency for the required output power remains challenging.

Variable-frequency, phase-shift/pulse-density modulation, or additional regulators for input/output regulation are widely used for the output control of wireless power transfer systems. Variable-frequency control may cause the loss of soft-switching due to gain-margin changes with load. It is also difficult to maintain soft switching under light-load conditions with phase-shift/pulse-density modulation. Additional converters are used as front-end or output-stage for impedance matching or regulation of the voltage/current at the load terminals. The use of additional power conversion stages ultimately degrades system efficiency and increases cost and size. All these methods may require a wireless communication link between the transmitter and receiver.

A desirable wireless energy transfer system can provide an optimized efficiency with soft switching of the inverter by receiver side master control. It does not require any additional converter stages or a wireless communication channel. Authors in [25] proposed a receiver stage semi-bridgeless active rectifier (S-BAR) to address these challenges as shown in Fig. 4.20. It comprises a half bridge resonant inverter with a coupling transformer and series resonant compensation, and a single-phase semi-bridge rectifier in the receiver side. The active rectifier is comprised of two transistors with anti-parallel diodes in the lower part of switching legs and two diodes in the upper parts.

S-BAR technology is a good option to maintain the compensation and regulation to the output power without any extra components on the receiver stage. However, misalignment conditions might limit the phase compensation due to boundary of duty regulation of S-BAR. In [26], an adaptively controlled variable switched capacitance is proposed to manage the load power under wide range misalignment variation for S-BAR inductive power transfer system. Maximum efficiency tracking system is presented to achieve constant current charging of super capacitor for

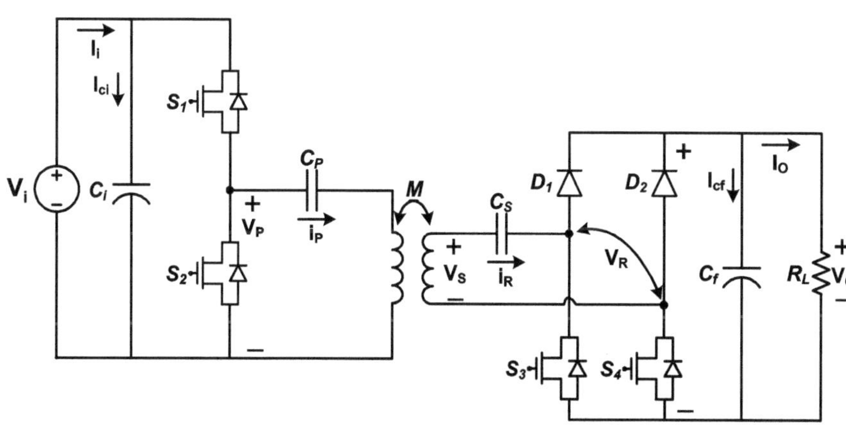

Fig. 4.20 A single phase semi bridgeless active rectifier (S-BAR)

S-BAR wireless energy transfer system in [27]. Also, an improved pulse density modulation of S-BAR wireless power transfer system is demonstrated for the multiple parallel receiver power control in [28].

A semi bridgeless active rectifier (S-BAR) circuit topology comprises a half bridge resonant inverter with an input dc source V_i, a coupler transformer with resonant capacitors C_P and C_S, and a single-phase semi-bridge rectifier in the secondary side for the wireless power transfer systems. The inverter is composed of MOSFETs S_1 and S_2, the resonant tank voltage V_P, and the resonant tank current I_P. The active rectifier is comprised of two transistors S_3 and S_4 with anti-parallel diodes in the lower part of switching legs and two diodes D_1 and D_2 in the upper parts. The phase-shift angle of the secondary side transistors regulates the output voltage V_O and power in the load R_L. The secondary side voltage is denoted here as V_S, the secondary side current is I_R, and the voltage across the rectifier tap center (receiver rectifier voltage) is V_R. The circuit analysis is simplified by assuming as,

- The diode is modeled as series combination of a diode forward voltage V_f and a diode forward resistance R_f during diode on state,
- The diode is an open circuit during diode off state,
- The diode lead inductance, junction capacitance, and charge lifetime are assumed zero,
- The rectifier input current i_R is driven pure sinusoidal source,
- The diode forward voltage V_f is lower than the rectifier input voltage amplitude V_{Rm},
- The output capacitor C_f is large enough for a constant dc output.

The receiver rectifier voltage v_R, current waveforms i_R , and primary side resonant tank voltage v_P are shown with different conduction angle β in Fig. 4.21. The phase controller angle α indicates the period of time between rising edge of rectifier voltage V_R and decreasing edge of primary resonant tank voltage V_P. This controller forms the rectifier conduction angle β and determines the output voltage V_O. The cases in Fig. 4.21a–f can occur at different β, the coupler transformer coupling coefficient factor K, and the load resistance R_L. In other words, even if α is kept constant, β varies with different K and R_L. It is important to mention that the converter operation should be avoided in some stages because of a very high output power. This is discussed in more detail in the following subsections. In this part, the phase shift operations are considered at a constant frequency f_s.

Controlling α and β is important to achieve various operating conditions. For $-\pi \leq \alpha \leq \pi$, the β is being changed as shown in Fig. 4.21. When the current i_R is leading the receiver rectifier voltage V_R as seen in Fig. 4.21a, b, c, and f, β increases and reaches the maximum value as depicted in Fig. 4.21c. At this moment, the active switches allow more current to pass, thus raising the output voltage. Another condition happens when the current is lagging the voltage after the maximum value point as shown in Fig. 4.21d and e. Decreasing the conduction angle β prevents current from passing to the output and thus lowers the output voltage. The other important thing at this cycle is that the β is decreasing but α is increasing. It means that the conduction angle can be in inverse or direct proportion to the controller

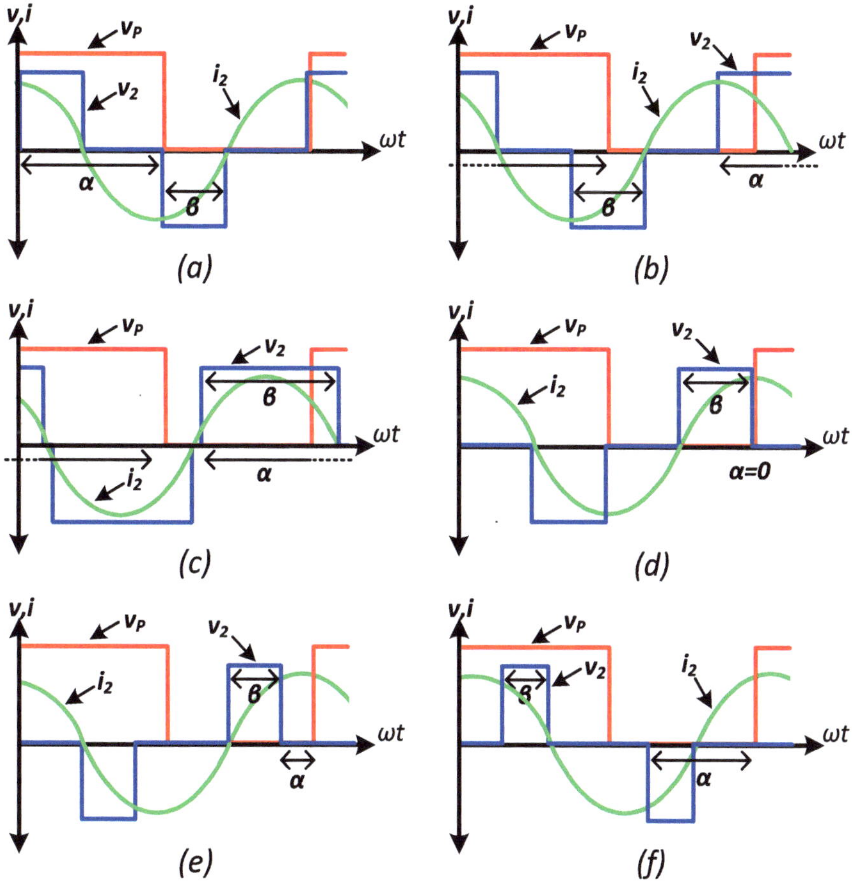

Fig. 4.21 Rectifier input voltage v_R and current i_R with respect to the primary resonant tank voltage v_P at different α and β: (**a**) $\alpha{=}-\pi$, $\beta{<}\alpha$; (**b**) $-\pi \leq \alpha \leq -\pi/2$, $\beta{<}\alpha$; (**c**) $-\pi/2 \leq \alpha \leq 0$, $\beta{<}\alpha$; (**d**) $\alpha{=}0$, $\beta{>}\alpha$; (**e**) $0 \leq \alpha \leq \pi/2$, $\beta{>}\alpha$; (**f**) $\pi/2 \leq \alpha \leq \pi$, $\beta{<}\alpha$

angle α. By varying the amount of time in which current reaches the output, the output voltage can be regulated depending on the control angle α and conduction angle β. It is important to explain that positive or negative sign of receiver rectifier voltage and current phase difference Θ indicates the load is inductive or capacitive; therefore, the reflected impedance to the receiver is shown as an inductive $\Theta > 0$, and as a capacitive $\Theta < 0$.

As seen in Fig. 4.21, there are three possible stages, $-V_O$, 0, V_O for the receiver side voltage v_R which is determined by the state of rectifier side switches. The circuit operation is described by considering inductive and capacitive load conditions in subsequent paragraphs.

For the inductive load:

$$v_R = \begin{cases} 0, & -\alpha < \omega t < (\pi - \alpha - \beta) \\ -V_O, & (\pi - \alpha - \beta) < \omega t < (\pi - \alpha) \\ 0, & (\pi - \alpha) < \omega t < (2\pi - \alpha - \beta) \\ V_O, & (2\pi - \alpha - \beta) < \omega t < (2\pi - \alpha) \end{cases} \tag{4.133}$$

The corresponding voltage $v_{R,1}$ value on the receiver terminals can be derived using the first harmonic approximation (FHA) as,

$$v_{R,1} = \frac{4V_O}{\pi} \sin\left(\frac{\beta}{2}\right) \sin\left(\omega t + \alpha + \frac{\beta}{2} - \frac{3\pi}{2}\right) d(\omega t) \tag{4.134}$$

The rectifier input current i_R can be also derived in a similar way with the above expression,

$$i_R = I_R \sin(\omega t + \alpha + \beta - 2\pi) d(\omega t) \tag{4.135}$$

The equivalent impedance $Z_{L,eq}$ is obtained by taking the ratio of $v_{R,1}$ and i_R in frequency domain, and is given by

$$Z_{L,eq} = \frac{4}{\pi^2} R_L (1 - cos\beta) \sin\left(\frac{\beta}{2}\right) e^{j\left[\frac{\pi}{2} - \frac{\beta}{2}\right]} \tag{4.136}$$

Taking real and imaginary portions of $Z_{L,eq}$, the equivalent resistance $R_{L,eq}$ and reactance $X_{L,eq}$ values are obtained as,

$$\begin{aligned} R_{L,eq} &= Re\left\{\tfrac{4}{\pi^2} R_L (1 - cos\beta) \sin\left(\tfrac{\beta}{2}\right) e^{j\left[\frac{\pi}{2} - \frac{\beta}{2}\right]}\right\} \quad 0 \le \beta \le \pi \\ X_{L,eq} &= Im\left\{\tfrac{4}{\pi^2} R_L (1 - cos\beta) \sin\left(\tfrac{\beta}{2}\right) e^{j\left[\frac{\pi}{2} - \frac{\beta}{2}\right]}\right\} \quad 0 \le \beta \le \pi \end{aligned} \tag{4.137}$$

$Z_{L,eq}$ is inductive, and load angle $(\pi/2\text{-}\beta/2)$ is always positive.

For the capacitive load:

$$v_R = \begin{cases} -V_O, & -\alpha < \omega t < (-\alpha + \beta) \\ 0, & (-\alpha + \beta) < \omega t < (\pi - \alpha) \\ V_O, & (\pi - \alpha) < \omega t < (\pi - \alpha + \beta) \\ 0, & (\pi - \alpha + \beta) < \omega t < (2\pi - \alpha) \end{cases} \tag{4.138}$$

FHA analysis results in the corresponding voltage $v_{R,1}$ value on the receiver terminals,

$$v_{R,1} = \frac{4V_O}{\pi} \sin\left(\frac{\beta}{2}\right) \sin\left(\omega t + \alpha - \frac{\beta}{2} - \frac{\pi}{2}\right) d(\omega t) \tag{4.139}$$

With the above expression, the rectifier input current i_R can be obtained as,

$$i_R = I_R \sin(\omega t + \alpha - \beta) d(\omega t) \tag{4.140}$$

The equivalent impedance $Z_{L,eq}$ can be defined as,

$$Z_{L,eq} = \frac{4}{\pi^2} R_L (1 - cos\beta) \sin\left(\frac{\beta}{2}\right) e^{j\left[\frac{\beta}{2} - \frac{\pi}{2}\right]} \tag{4.141}$$

The equivalent resistance $R_{L,eq}$ and reactance $X_{L,eq}$ values are acquired from the above equation as,

$$\begin{aligned} R_{L,eq} &= Re\left\{\tfrac{4}{\pi^2} R_L (1 - cos\beta) \sin\left(\tfrac{\beta}{2}\right) e^{j\left[\frac{\beta}{2} - \frac{\pi}{2}\right]}\right\} - \pi \le \beta \le 0 \\ X_{L,eq} &= Im\left\{\tfrac{4}{\pi^2} R_L (1 - cos\beta) \sin\left(\tfrac{\beta}{2}\right) e^{j\left[\frac{\beta}{2} - \frac{\pi}{2}\right]}\right\} - \pi \le \beta \le 0 \end{aligned} \tag{4.142}$$

The load angle *(β/2-π/2)* is always negative, and $Z_{L,eq}$ is capacitive. Figure 4.22 shows the load equivalent resistance $R_{L,eq}$ and reactance $X_{L,eq}$ as a function of β and load resistance R_L.

The output voltage V_O is held via the secondary rectifier switches and diodes as long as the current i_R is continuous. Thus, the output current I_O can be described by integrating the V_O/R_L ratio over the 0 to β interval.

$$I_O = \frac{V_O}{R_L} = \frac{1}{\pi} \int_0^{\beta} i_R \sin(\omega t) d\omega t \tag{4.143}$$

This result gives the maximum rectifier input current I_R as a function of β.

$$I_R = \frac{V_O \pi}{R_L\,(1 - \cos(\beta))} = \frac{I_O \pi}{(1 - \cos(\beta))} \tag{4.144}$$

The rectifier current transfer function M_{IR} can be found as,

$$M_{IR} = \frac{(1 - \cos(\beta))}{\pi} \tag{4.145}$$

The rectifier voltage v_R and current i_R waveforms at $-90 > \beta > -180$ are important for the switch ZVS/ZCS conditions. When the system is operated in the

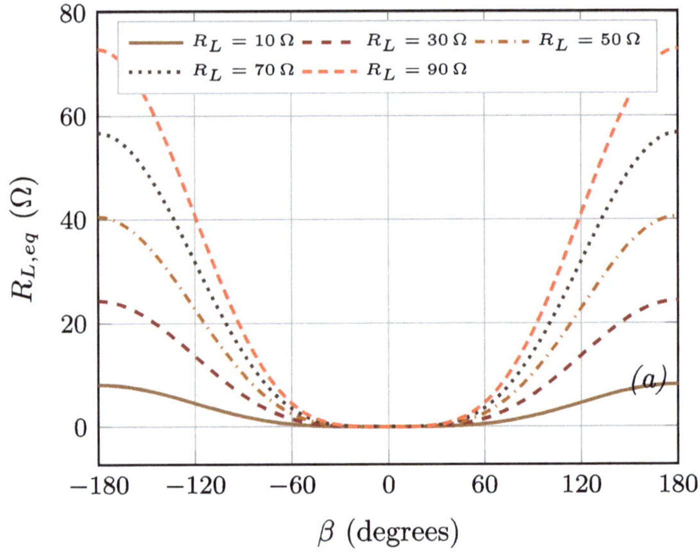

(a)

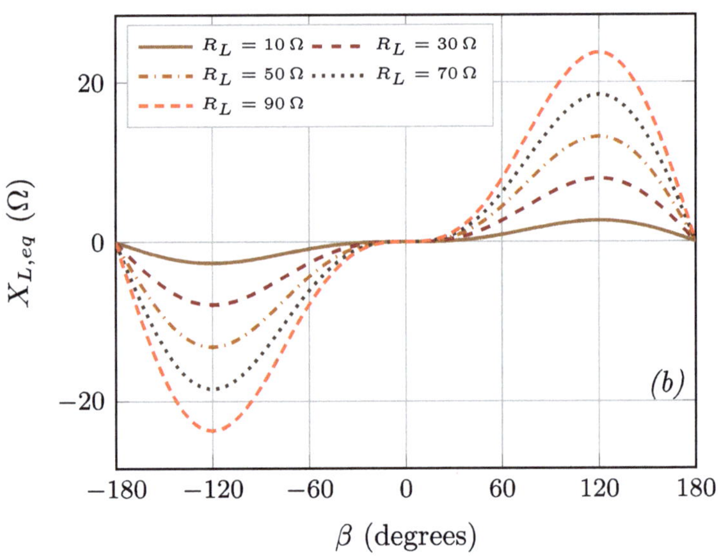

(b)

Fig. 4.22 The equivalent load impedance $Z_{L,eq}$ variation with conduction angle β and load resistance R_L: (**a**) equivalent resistance $R_{L,eq}$; (**b**) equivalent reactance $X_{L,eq}$

ZVS region, the secondary side current i_R is leading the secondary side voltage v_R and the synchronous MOSFETs S_3 and S_4 achieve the ZVS condition. Before the gate signal is turned on, the current in the circuit is conducting through the parallel connected diodes. Hence, the switch voltage drops to zero at this condition. The second stage is also crucial when $90 < \beta < 180$. The system is operated in ZVS region from the primary side, while the secondary side is in ZCS region, i_R is lagging the secondary voltage v_R unlike the first condition. Even under the worst circumstances, achievable ZVS process is one of the most important issues in order to reduce switching losses which improves the converter performance and efficiency. Due to reverse recovery process at ZCS, the ZVS operation provides fewer losses in the converter. Thus, the system operation is preferred between $-90 > \beta > -180$ in order to achieve the ZVS region in the secondary side MOSFETs S_3 and S_4.

The gate drive losses P_g can be calculated using the equation of (4.81). The rectifier MOSFETs S_3 and S_4 are not experiencing turn–on losses when $\beta<0$, but there are turn–off losses. Each MOSFET turn-off loss $P_{t\text{-}off}$ can be approximated as,

$$P_{t-off} = f_s V_R \left| I_R \sin(\beta - \pi) \right| \left(\frac{2t_r}{3} + t_f \right) \tag{4.146}$$

where β is the conduction angle defined above. As long as the receiver side active switches are operated in the ZVS region, MOSFETs freewheeling diodes are turning-off at ZVS and lossless. Also, turn-on losses are negligible. Moreover, upper diode does not impose switching losses at the turning-on and off because of the zero current switching. Each MOSFET conduction loss is,

$$P_{DS} = \frac{V_F I_R (1 + cos\beta)}{\pi} + \frac{I_R^2 r_{\text{DSon}}}{2} \tag{4.147}$$

where V_F is the MOSFET internal diode forward voltage drop. r_{DSon} shows the MOSFET internal resistance. Each upper diode conduction loss P_D and output filter capacitor power loss P_{rC} can be obtained as,

$$P_D = \frac{V_f I_R (1 - cos\beta)}{\pi} \tag{4.148}$$

$$P_{rC} = r_C I_R^2 \left[\left(\frac{\beta}{2} - \frac{\sin(2\beta)}{4} \right) \frac{1}{4} - \left(\frac{1 - cos\beta}{\pi} \right)^2 \right] \tag{4.149}$$

where V_f is the upper diode forward voltage, and r_C is the output capacitor ESR. The total rectifier losses P_T can be obtained as,

$$\begin{aligned} P_T &= 2P_g + 2P_{t-off} + 2P_{DS} + 2P_D + P_{rLC} \\ &= 2Q_g V_{\text{GSpp}} f_s + 2V_R \left| I_R \sin(\beta - \pi) \right| f_s \left(\frac{2t_r}{3} + t_f \right) \end{aligned}$$

$$+\frac{2V_F I_R(1+\cos\beta)}{\pi}+I_R^2 r_{\text{DSon}}+\frac{2V_f I_R(1-\cos\beta)}{\pi}$$
$$+r_C I_R^2\left[\left(\frac{\beta}{2}-\frac{\sin(2\beta)}{4}\right)\frac{1}{4}-\left(\frac{1-\cos\beta}{\pi}\right)^2\right] \tag{4.150}$$

In order to calculate the efficiency η_R of the circuit, equations for the total losses P_T and the output power are needed. The output power P_O can be calculated based on the secondary side current I_R as,

$$P_O=\frac{|I_R|^2 R_{L,eq}}{2} \tag{4.151}$$

The overall efficiency η_R of the system can be estimated by

$$\begin{aligned}\eta_R&=\frac{1}{1+\frac{P_T}{P_O}}\\&=\frac{1}{1+\frac{2Q_g V_{\text{GSpp}} f_s}{P_O}+\frac{2V_R|I_R\sin(\beta-\pi)|f_s\left(\frac{2t_r}{3}+t_f\right)}{P_O}+\frac{4V_F(1+\cos\beta)}{\pi R_{L,eq}}}\\&\quad\times\frac{1}{\frac{2r_{\text{DSon}}}{R_{L,eq}}+\frac{4V_f(1-\cos\beta)}{\pi R_{L,eq}}+\frac{2r_C}{R_{L,eq}}\left[\left(\frac{\beta}{2}-\frac{\sin(2\beta)}{4}\right)\frac{1}{4}-\left(\frac{1-\cos\beta}{\pi}\right)^2\right]}\end{aligned} \tag{4.152}$$

The voltage transfer function M_V of the rectifier is

$$M_{VR}=\frac{V_O}{V_R}=M_{IR}\frac{R_L}{R_{L,eq}\eta_R} \tag{4.153}$$

The S-BAR converter topology has reduced number of switches on the secondary side as compared to typical inductive energy transfer secondary side DC–DC converter topologies. The output voltage or current can be regulated using a controller only in the secondary side. With this feature, S-BAR can be an excellent solution for wireless energy transfer applications. Also, the concepts can be used in multiple output receiver wireless applications since the output control can be regulated only in the receiver side. The proposed receiver topology utilization can also be facilitated for the impedance matching protection by adjusting reflected equivalent resistance value.

4.3 DC–DC Converters

After the high-frequency ac-dc conversion stage discussed in Sect. 4.2, many wireless power transfer (WPT) receivers employ an additional DC–DC converter to regulate the voltage or current delivered to the battery, dc bus, or electronic load. This post-rectifier stage is often required because the dc output of the rectifier varies with coupling coefficient, alignment, coil separation, resonant tuning, operating frequency, and load. In battery-oriented receivers, this stage also serves as the natural location for implementing constant-current and constant-voltage charging, current limiting, and bus stabilization [6, 8, 29]. Recent receiver-side reviews and charging-oriented studies further confirm that the rectifier-plus-DC–DC architecture remains highly relevant when wide regulation range, charging-profile enforcement, and tolerance to coupling variation are required [30, 31].

The converters discussed in this section utilize PWM to control the output. The PWM signal is generated by a switch, for instance a transistor, which is turned on for time t_{on} and then turned off for time t_{off}. This on/off process is repeated continuously with a period $T = t_{on} + t_{off}$. Hence, the switch duty ratio D can be defined as

$$D = \frac{t_{on}}{t_{on} + t_{off}} \tag{4.154}$$

From the above equation, it is evident that the theoretical range of duty ratio D is from zero to one. In the following idealized discussion, V_{IN} denotes the dc voltage available at the converter input, which in a WPT receiver is usually the rectifier output voltage, while V_{OUT} denotes the regulated voltage applied to the load. The ideal conversion-ratio relations presented in this section are the standard continuous-conduction-mode (CCM) results for PWM converters and serve as an initial design step before parasitic elements and switching losses are considered [32].

4.3.1 *Step-Down Converter*

A step-down DC–DC converter, commonly known as a buck converter, is frequently used on the secondary side of a WPT system when the rectified voltage is higher than the required battery or load voltage over the expected operating range. This situation is common when the magnetic coupling is relatively strong and the rectifier produces a dc bus that must be reduced to a lower regulated level. In consumer, biomedical, and vehicular WPT receivers, the buck stage is attractive because it provides simple control, a low device count, and a continuous inductor output current that is well suited to downstream battery and dc-bus filtering requirements [6, 33].

Figure 4.23 shows the buck converter topology. It consists of a dc input voltage source V_{IN}, controlled switch S, diode D, filter inductor L, filter capacitor C, and

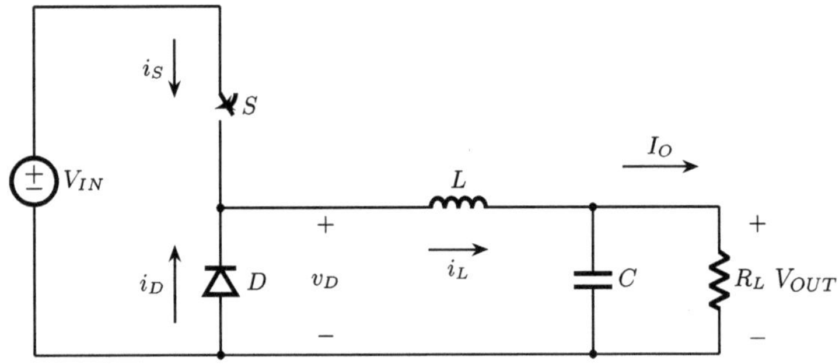

Fig. 4.23 The buck converter topology diagram

load resistance R_L. Typical converter waveforms are shown in Fig. 4.24 under the assumption that the inductor current is always positive. When the switch S is driven to the on state, the diode D is reverse biased.

When the switch S is off, the diode conducts to maintain uninterrupted current in the inductor. The relationship among the input voltage, output voltage, and the switch duty ratio D can be derived from the waveform of the voltage across the inductor L. The voltage across the inductor is equal to $V_{\mathrm{IN}} - V_{\mathrm{OUT}}$ when the switch S is on and equal to $-V_{\mathrm{OUT}}$ when the switch is off. According to Faraday's law, the inductor volt-second product over a period of steady-state operation is zero. For the buck converter,

$$(V_{\mathrm{IN}} - V_{\mathrm{OUT}})\,DT = V_{\mathrm{OUT}}(1 - D)T \tag{4.155}$$

Hence, the dc voltage transfer function, defined as the ratio of the output voltage to the input voltage, is

$$M_V = \frac{V_{\mathrm{OUT}}}{V_{\mathrm{IN}}} = D \tag{4.156}$$

From the above equation, it follows that the output voltage of the buck converter lies within the range from 0 to V_{IN}.

The filter inductor current i_L consists of a dc component I_O with a triangular ac component. This component represents the inductor current ripple. Its peak-to-peak value can be calculated as

$$I_{Lr} = \frac{V_{\mathrm{OUT}} t_{\mathrm{off}}}{L} = \frac{V_{\mathrm{OUT}}(1 - D)T}{L} \tag{4.157}$$

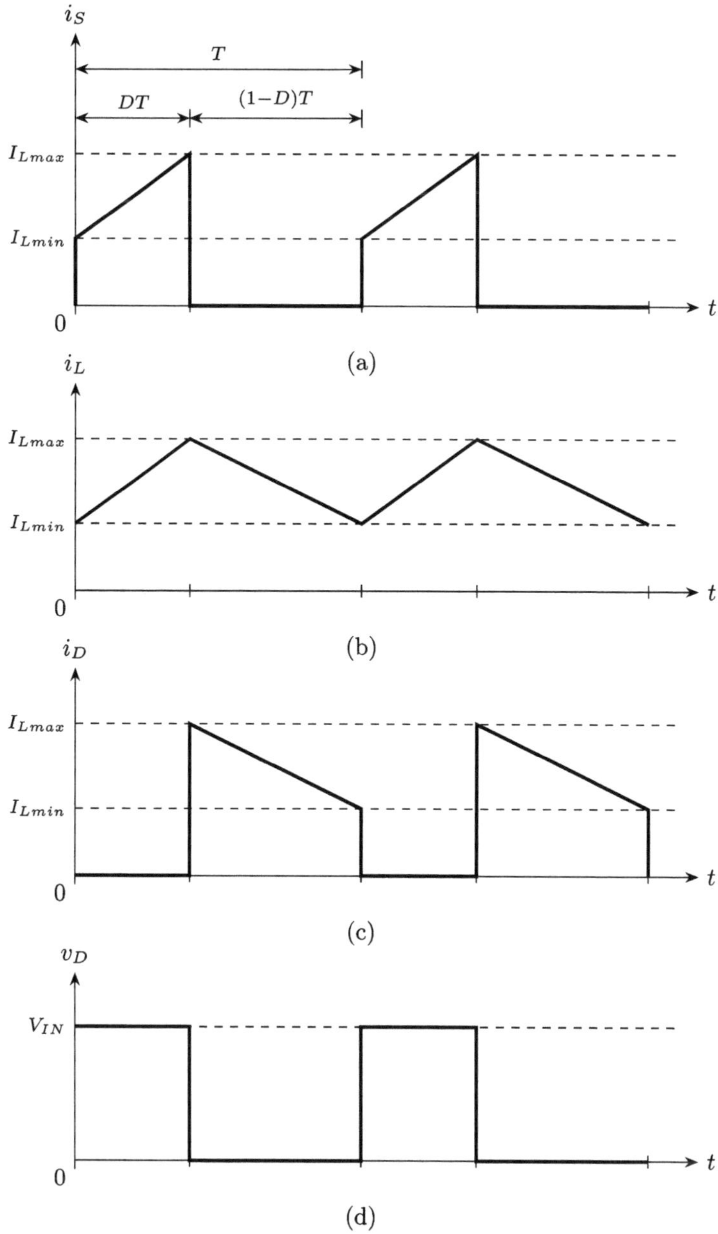

Fig. 4.24 Typical voltage and current waveforms of the buck converter: (**a**) switch current, (**b**) inductor current, (**c**) diode current, (**d**) diode voltage

The maximum and minimum values of the inductor current i_L are

$$I_{L\max} = I_O + \frac{V_{OUT}(1-D)T}{2L} \tag{4.158}$$

$$I_{L\min} = I_O - \frac{V_{OUT}(1-D)T}{2L} \tag{4.159}$$

In most design cases, the filter inductance L is selected to keep the current ripple below a certain limit $I_{Lr\max}$. Thus, the inductance value can be calculated as

$$L_{\min} = \frac{V_{OUT}(1-D)T}{I_{Lr\max}} \tag{4.160}$$

Almost the entire ac component of the inductor current flows through the filter capacitor as current i_C. Current i_C causes a small voltage ripple across the dc output voltage V_{OUT}. To limit the peak-to-peak value of the ripple voltage below a certain value V_r, the filter capacitance C must be greater than

$$C_{\min} = \frac{V_{OUT}(1-D)}{8V_r L f^2} = \frac{I_{Lr\max}}{8V_r f} \tag{4.161}$$

Equations (4.160) and (4.161) are the key design equations for the buck converter. The input and output dc voltages, the duty ratio D, and the range of load resistance R_L are usually determined by preliminary specifications. The designer must determine the values of the passive components L and C, and the switching frequency f. In WPT applications, the buck converter is attractive when the pickup voltage at nominal or strong coupling is higher than the required battery or dc-bus voltage and a compact, high-efficiency post-regulator is desired. The converter should be designed so that the minimum inductor current remains positive over the intended load range; otherwise, the converter enters discontinuous conduction mode and its conversion ratio becomes load dependent. Thus, the CCM boundary condition $I_O = I_{Lr}/2$ serves as an important design check in addition to the ripple limits given by (4.160) and (4.161) [32]. In practice, the ripple limits need to be considered with respect to the presence and value of the equivalent series resistance (ESR) of the filter capacitor, which is an important parasitic component in any DC–DC converter. For sufficiently large capacitance, specifically when $C > \max\{(1-D)/(2r_C f), D/(2r_C f)\}$, the peak-to-peak output-voltage ripple is determined primarily by the capacitor ESR and may be written as

$$V_r = r_C I_{Lr} = \frac{r_C V_{OUT}(1-D)}{Lf} \tag{4.162}$$

Thus, once this condition is met, a further increase of C alone does not significantly reduce the ripple unless r_C is also reduced [34].

4.3.2 Step-up Converter

A step-up converter, or PWM boost converter, is useful in WPT receivers when the rectifier output is lower than the required regulated voltage. This may occur under weak coupling, increased air gap, coil misalignment, or intentionally reduced link voltage. By stepping up the post-rectifier voltage, the converter can extend the operating range of the receiver while maintaining the required dc output. Such operating conditions are common in variable-coupling WPT systems, including vehicular chargers and loosely coupled consumer platforms [4, 6].

Figure 4.25 presents the boost converter topology. It is composed of a dc input voltage source V_{IN}, boost inductor L, controlled switch S, diode D, filter capacitor C, and load resistance R_L. Typical converter waveforms are shown in Fig. 4.26 under the assumption that the inductor current is always positive. When the switch S is in the on state, the current in the boost inductor increases linearly and the diode D is off. When the switch S is turned off, the energy stored in the inductor is released through the diode D to the output RC network. The voltage across the inductor is equal to V_{IN} when the switch S is on and equal to $V_{\text{IN}} - V_{\text{OUT}}$ when the switch is off. Using Faraday's law for the boost inductor,

$$V_{\text{IN}} DT = (V_{\text{OUT}} - V_{\text{IN}})\,(1 - D)T \tag{4.163}$$

from which the dc voltage transfer function can be derived as

$$M_V = \frac{V_{\text{OUT}}}{V_{\text{IN}}} = \frac{1}{1 - D} \tag{4.164}$$

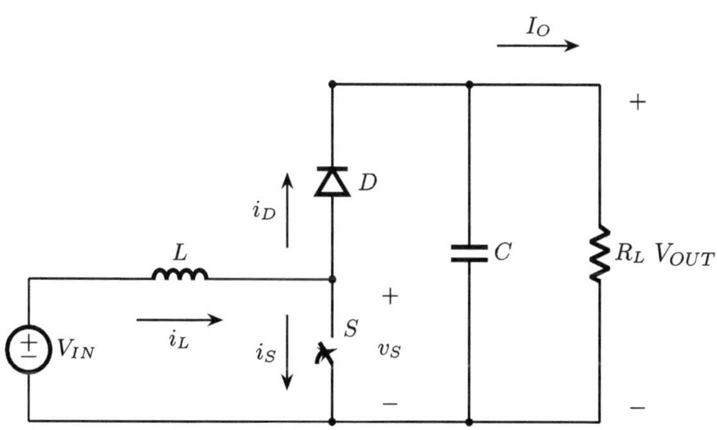

Fig. 4.25 The boost converter topology diagram

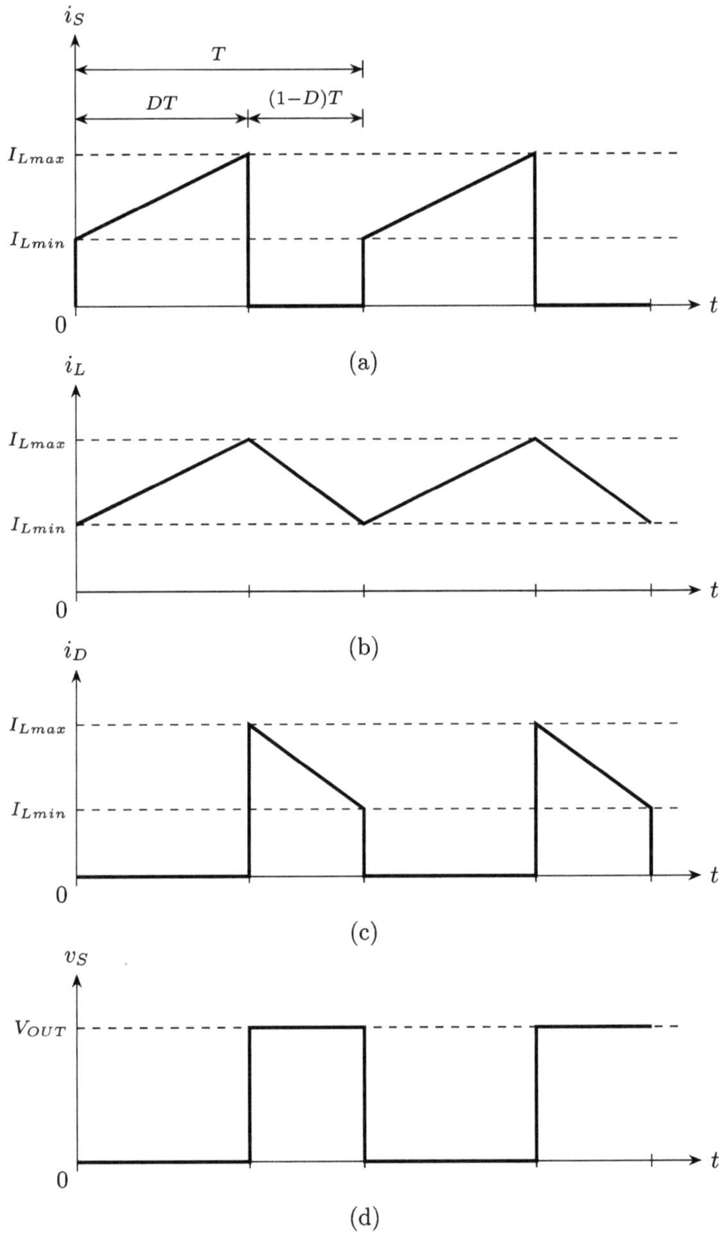

Fig. 4.26 Typical voltage and current waveforms of the boost converter: (**a**) switch current, (**b**) inductor current, (**c**) diode current, (**d**) switch voltage

From (4.164), it follows that the output voltage of the ideal boost converter lies within the range from V_{IN} to infinity. The step-up ratio of a practical boost converter is limited to about 10 due to the increase in losses with the duty ratio. Still, duty-ratio limitation and output over-voltage protection are strongly recommended because the boost converter can generate very high output voltage levels.

The peak-to-peak value of the inductor current ripple can be calculated as

$$I_{Lr} = \frac{V_{\mathrm{IN}} t_{\mathrm{on}}}{L} = \frac{V_{\mathrm{IN}} DT}{L} \tag{4.165}$$

The maximum and minimum values of the inductor current i_L are

$$I_{L\,\max} = I_O + \frac{V_{\mathrm{IN}} DT}{2L} \tag{4.166}$$

$$I_{L\,\min} = I_O - \frac{V_{\mathrm{IN}} DT}{2L} \tag{4.167}$$

In most cases, the filter inductance L is selected to keep the current ripple below a certain limit $I_{Lr\,\max}$. Thus, the required inductor value can be calculated as

$$L_{\min} = \frac{V_{\mathrm{IN}} DT}{I_{Lr\,\max}} \tag{4.168}$$

In the boost converter, current is delivered to the output RC network only when diode D is conducting. Thus, this current is discontinuous, which implies a larger filter-capacitor requirement compared to buck-derived converters in order to limit the output voltage ripple. The filter capacitor must supply the output dc current to the load when the diode D is not conducting. The minimum value of the filter capacitance that results in the voltage ripple V_r is given by

$$C_{\min} = \frac{V_{\mathrm{OUT}} D}{V_r R_L f} \tag{4.169}$$

Similarly to the buck converter, the output-voltage ripple contribution due to the output-capacitor ESR should also be taken into account. Unlike in the buck converter, this ESR-related ripple component adds to the capacitance-induced ripple [32].

In a WPT receiver, the boost stage can be used to maintain the required dc bus voltage under weak-coupling or large-air-gap conditions. This benefit must be balanced against the higher current stress, the larger output-capacitor requirement, and the possibility of excessive switch-voltage stress at high duty ratio. Consequently, boost post-regulators are most attractive when the desired operating envelope cannot be covered by retuning alone and when the control system can enforce safe duty-ratio and over-voltage protection limits [8, 33].

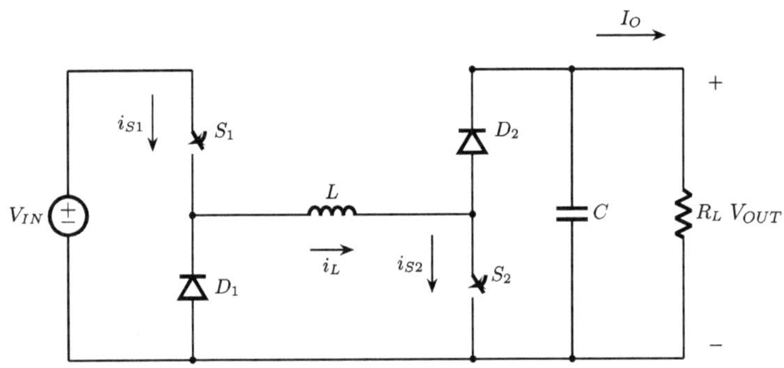

Fig. 4.27 The integrated buck and boost converter topology diagram

4.3.3 Converter Choices for WPT Receivers

Figure 4.27 depicts an integrated buck-boost converter. It consists of a dc input voltage source V_{IN}, inductor L, two controlled switches S_1 and S_2, two diodes D_1 and D_2, filter capacitor C, and load resistance R_L. This converter has two operating modes: buck and boost. In the buck mode, switch S_2 is off and switch S_1 is turned on and off with duty ratio D and frequency f. Assuming a positive current in the inductor L, diode D_2 continuously conducts in this operation mode. Thus, the converter operates as a conventional buck converter and all related equations are applicable.

In the second operating mode, the converter behaves as a boost converter. In this case, switch S_1 is kept continuously on and diode D_1 does not conduct. Switch S_2 is turned on and off with duty ratio D and frequency f. Thus, the converter operates as a conventional boost converter, and all related equations are applicable.

The integrated buck-boost converter is especially useful in WPT systems in which the rectifier output may be above the target load voltage under nominal alignment and below it under weak coupling or misalignment. Such a wide regulation range is not achievable with a buck or boost converter alone. It can also be achieved by SEPIC or Ćuk converters, but those are higher-order converters and their analysis and control are more complex. For receivers exposed to large coupling variation, the noninverting buck-boost family offers a practical compromise between regulation range, passive-component count, and control complexity [6, 33].

Some WPT applications, particularly battery-powered receivers and energy-storage interfaces, require bidirectional power-flow capability in the dc stage. In non-isolated converters, this can be achieved by applying synchronous rectification and appropriate control methods. Figure 4.28 shows an example of such a converter. After replacing a diode with a controlled switch, the buck and boost converter topologies become structurally identical, and the direction of power flow is determined by the control strategy. Thus, the converter can operate in one direction as a buck and in the reverse direction as a boost converter. The integrated buck-boost

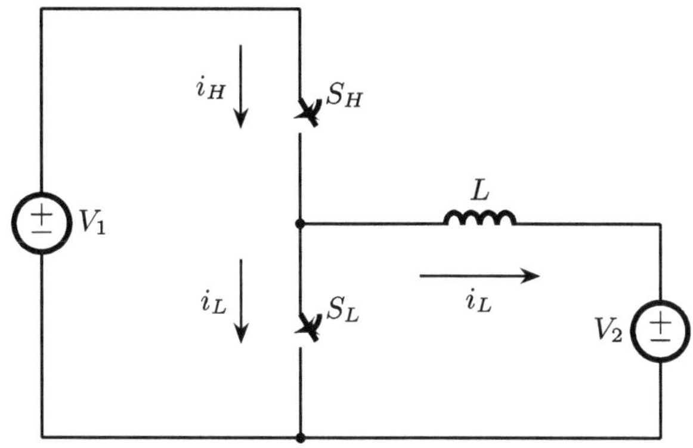

Fig. 4.28 The bidirectional non-isolated converter topology diagram

topology provides greater flexibility in bidirectional operation with respect to the input/output voltage ratio. Bidirectional capability is especially relevant in advanced battery interfaces and vehicle-to-grid (V2G) platforms, where the same hardware may be expected to support charging and reverse energy flow [29].

For applications where additional galvanic isolation or very high power capability is required, the dual active bridge (DAB) converter can be used. In many WPT systems, the magnetic link already provides the main isolation barrier, and therefore a non-isolated post-rectifier converter is often preferred for simplicity and efficiency. Nevertheless, DAB-based converters remain attractive when a wide conversion ratio, soft-switching operation, bidirectional power flow, or modular interfaces are required [35].

4.3.4 Reflected-Load Control and Practical Issues

In a WPT receiver, the DC–DC converter cannot be designed independently of the resonant link and rectifier. The converter duty ratio determines the equivalent resistance seen at its input, and this quantity is reflected through the rectifier to the resonant network. Consequently, converter control affects not only dc-output regulation but also the operating point, current stress, soft-switching conditions, and efficiency of the overall wireless link. This observation is fully consistent with Sect. 4.1.1, where cascaded DC–DC stages are used to modify the effective load presented to the pickup circuit [1, 2]. Recent analyses for SS-compensated WPT receivers also indicate that the controller dynamics of the DC–DC stage must be coordinated with the resonant interface rather than treated as a stand-alone downstream regulator [36].

Practical design must also account for parasitic elements and non-ideal devices. Inductor copper loss, capacitor equivalent series resistance (ESR), diode forward drop, MOSFET on-state resistance, reverse recovery, dead time, gate-drive loss, switching loss, and layout-induced ringing all reduce efficiency and may alter the ideal relations given in (4.154)–(4.171). The switching frequency should be high enough to reduce passive component size, yet low enough to satisfy thermal, efficiency, and electromagnetic-interference (EMI) constraints. In WPT receivers, another important issue is the interaction between the post-regulator bandwidth and the slower tuning or primary-side control loops; poorly coordinated control can produce bus oscillation, loss of zero-voltage switching, or excessive circulating current. Therefore, the final choice of post-rectifier converter topology is dictated not only by the required dc voltage conversion ratio, but also by coupling variation, allowable reflected-load range, battery-charging profile, semiconductor stress, and the overall control strategy of the wireless power transfer system [6, 8, 33]. At the same time, recent direct-charging investigations show that some regulation tasks can be shifted toward the primary side, which provides a useful comparison benchmark for deciding how much functionality should remain in the receiver-side DC–DC converter [37, 38].

For the ideal CCM buck and boost converters, the input resistance presented to the rectifier can be written directly from the conversion ratio and power-balance relations. These expressions are useful in WPT because they show how duty-ratio control electronically transforms the effective load seen by the resonant link [1, 33].

$$R_{\mathrm{IN}} = \frac{V_{\mathrm{IN}}}{I_{\mathrm{IN}}} = \frac{R_L}{D^2} \tag{4.170}$$

$$R_{\mathrm{IN}} = \frac{V_{\mathrm{IN}}}{I_{\mathrm{IN}}} = R_L(1-D)^2 \tag{4.171}$$

Equation (4.170) indicates that the buck converter increases the resistance seen at its input as the duty ratio decreases, while (4.171) shows that the boost converter reduces the effective input resistance as the duty ratio increases. Hence, the post-rectifier converter is not merely a local dc voltage regulator; it is also an impedance-transforming control element that influences resonant current, transferred power, and overall efficiency. This viewpoint is particularly important in receivers that must sustain good performance over large coil displacement or a wide battery-voltage range [4, 6].

References

1. J. Shin, F. Wu, D. Czarkowski, M.T. Outeiro, PWM controllable inductance using a Class E rectifier for real-time resonance tuning, in *IECON 2016 – 42nd Annual Conference of the IEEE Industrial Electronics Society* (2016), pp. 5645–5650. https://doi.org/10.1109/iecon.2016.7793987
2. J. Shin, M.T. Outeiro, D. Czarkowski, New real-time tuning method for wireless power transfer systems, in *2016 IEEE Wireless Power Transfer Conference (WPTC)*, Aveiro (2016), pp. 1–4. https://doi.org/10.1109/wpt.2016.7498798
3. Y. Zheng, F. Feng, X. Wang, Y. Huang, P. Xie, Power stability enhancement in inductive wireless power transfer systems: a review. IEEE Access (2024). https://doi.org/10.1109/access.2024.3467099

4. S.Y.R. Hui, W. Zhong, C.K. Lee, A critical review of recent progress in mid-range wireless power transfer. IEEE Trans. Power Electron. **29**(9), 4500–4511 (2014). https://doi.org/10.1109/tpel.2013.2249670
5. M.P. Theodoridis, Effective capacitive power transfer. IEEE Trans. Power Electron. **27**(12), 4906–4913 (2012). https://doi.org/10.1109/tpel.2012.2192502
6. O.C. Onar, M. Chinthavali, S.L. Campbell, L.E. Seiber, C.P. White, Vehicular integration of wireless power transfer systems and hardware interoperability case studies. IEEE Trans. Ind. Appl. **55**(5), 5223–5234 (2019). https://doi.org/10.1109/tia.2019.2928482
7. E. Asa, K. Colak, D. Czarkowski, Analysis of multi-output half- wave semi-synchronous rectifier with a uniform magnetic field transmitter, in *2016 IEEE Applied Power Electronics Conference and Exposition (APEC)*, Long Beach (2016), pp. 2587–2591. https://doi.org/10.1109/apec.2016.7468229
8. V. Shevchenko, O. Husev, R. Strzelecki, B. Pakhaliuk, N. Poliakov, N. Strzelecka, Compensation topologies in IPT systems: standards, requirements, classification, analysis, comparison and application. IEEE Access **7**, 120559–120580 (2019). https://doi.org/10.1109/access.2019.2937891
9. K. Woronowicz, A. Safaee, T.R. Dickson, Single-phase zero reactive power wireless power transfer topologies based on Boucherot bridge circuit concept. Can. J. Electrical Comput. Eng. **38**(4), 323–337 (2015). https://doi.org/10.1109/cjece.2015.2468057
10. L. Nela, G. Kampitsis, J. Ma, E. Matioli, Fast-switching tri-anode Schottky barrier diodes for monolithically integrated GaN-on-Si power circuits. IEEE Electron Device Lett. **41**(1), 99–102 (2020). https://doi.org/10.1109/led.2019.2957700
11. S. Han, S. Yang, R. Li, X. Wu, K. Sheng, Current collapse free and fast reverse recovery performance in vertical GaN-on-GaN Schottky barrier diode. IEEE Trans. Power Electron. **34**(6), 5012–5018 (2019). https://doi.org/10.1109/tpel.2018.2876444
12. M. Adamowicz, S. Giziewski, J. Pietryka, Z. Krzeminski, Performance comparison of SiC Schottky diodes and silicon ultra fast recovery diodes, in *7th International Conference-Workshop Compatibility and Power Electronics (CPE)*, Tallinn (2011), pp. 144–149. https://doi.org/10.1109/cpe.2011.5942222
13. M. Tsukuda, H. Imaki, I. Omura, Ultra-fast lateral 600 V silicon PiN diode superior to SiC-SBD, in *2014 IEEE 26th International Symposium on Power Semiconductor Devices & IC's (ISPSD)*, Waikoloa (2014), pp. 31–34. https://doi.org/10.1109/ispsd.2014.6855968
14. V. Kindl, M. Zavrel, M. Tyrpekl, M. Frivaldsky, J. Skorvaga, Wireless power transfer system with current-doubler rectifier on the secondary side—analysis, modeling, and verification. Electronics **12**(23), 4818 (2023). https://doi.org/10.3390/electronics12234818
15. M.K. Kazimierczuk, D. Czarkowski, Resonant Power Converters, 2nd ed. (Wiley, Hoboken, 2011). ISBN: 978-0-470-90538-8
16. M.K. Kazimierczuk, Class D current-driven rectifiers for resonant DC/DC converter applications. IEEE Trans. Ind. Electron. **38**(5), 344–354 (1991). https://doi.org/10.1109/41.97554
17. C. Ekkaravarodome, V. Chunkag, K. Jirasereeamornkul, M.K. Kazimierczuk, Class-D zero-current-switching rectifier as power-factor corrector for lighting applications. IEEE Trans. Power Electron. **29**(9), 4938–4948 (2014). https://doi.org/10.1109/tpel.2013.2284872
18. S. Mangkalajan, C. Ekkaravarodome, K. Jirasereeamornkul, P. Thounthong, K. Higuchi, M.K. Kazimierczuk, A single-stage LED driver based on ZCDS Class-E current-driven rectifier as a PFC for street-lighting applications. IEEE Trans. Power Electron. **33**(10), 8710–8727 (2018). https://doi.org/10.1109/tpel.2017.2780088
19. T. Suetsugu, M.K. Kazimierczuk, Integrated Class DE synchronized DC-DC converter for on-chip power supplies, in *37th IEEE Power Electronics Specialists Conference*, Jeju (2006), pp. 1–5. https://doi.org/10.1109/pesc.2006.1711803
20. J.-W. Yang, H.-L. Do, Soft-switching dual-flyback DC-DC converter with improved efficiency and reduced output ripple current. IEEE Trans. Ind. Electron. **64**(5), 3587–3594 (2017). https://doi.org/10.1109/tie.2017.2652404
21. C. Fei, Q. Li, F.C. Lee, Digital implementation of adaptive synchronous rectifier (SR) driving scheme for high-frequency LLC converters with microcontroller. IEEE Trans. Power Electron. **33**(6), 5351–5361 (2018). https://doi.org/10.1109/tpel.2017.2731942

22. M.K. Kazimierczuk, W. Szaraniec, S. Wang, Analysis and design of parallel resonant converter with high Q_L. IEEE Trans. Aerosp. Electron. Syst. **28**(1), 35–50 (1992). https://doi.org/10.1109/7.135431
23. A. Fernandez, D. Lamar, M. Rodriguez, M.M. Hernando, J. Sebastian, M. Arias, Self-driven synchronous rectification system with input voltage tracking for converters with a symmetrically driven transformer. IEEE Trans. Ind. Electron. **56**(5), 1440–1445 (2009). https://doi.org/10.1109/tie.2009.2012455
24. A.H. Asghari, H. Farzanehfard, Synchronous rectifier driving circuit based on voltage-driven method with energy recovery. IET Power Electron. **7**(4), 765–774 (2014). https://doi.org/10.1049/iet-pel.2013.0279
25. K. Colak, E. Asa, M. Bojarski, D. Czarkowski, O.C. Onar, A novel phase shift control of semibridgeless active rectifier for wireless power transfer. IEEE Trans. Power Electron. **30**(11), 6288–6297 (2015). https://doi.org/10.1109/tpel.2015.2430832
26. E.S. Lee, B.G. Choi, J.S. Choi, D.T. Nguyen, C.T. Rim, Wide range adaptive IPT using dipole coils with a reflector by variable switched capacitance. IEEE Trans. Power Electron. **32**(10), 8054–8070 (2017). https://doi.org/10.1109/tpel.2016.2637931
27. Z. Li, K. Song, J. Jiang, C. Zhu, Constant current charging and maximum efficiency tracking control scheme for supercapacitor wireless charging. IEEE Trans. Power Electron. **33**(10), 9088–9100 (2018). https://doi.org/10.1109/tpel.2018.2793312
28. M. Fan, L. Shi, Z. Yin, L. Jiang, F. Zhang, Improved pulse density modulation for semi-bridgeless active rectifier in inductive power transfer system. IEEE Trans. Power Electron. **34**(6), 5893–5902 (2019). https://doi.org/10.1109/tpel.2018.2867902
29. M. Yilmaz, P.T. Krein, Review of battery charger topologies, charging power levels, and infrastructure for plug-in electric and hybrid vehicles. IEEE Trans. Power Electron. **28**(5), 2151–2169 (2013). https://doi.org/10.1109/tpel.2012.2212917
30. J. Zhang, L. Kong, Z. Wang, Y. Wang, Y. Liu, X. Gao, C. Zhu, Receiver-side topologies for wireless power transfer systems: a comprehensive review of the design, challenges, and future trends. Energies **18**(6), 1493 (2025). https://doi.org/10.3390/en18061493
31. D. Purushothaman, R. Narayanamoorthi, V.K. Ramachandramurthy, Charging ahead: unlocking the potential of constant voltage and constant current modes in WPT for EVs. J. Energy Storage **96**, 112603 (2024). https://doi.org/10.1016/j.est.2024.112603
32. M.K. Kazimierczuk, *Pulse-Width Modulated DC–DC Power Converters*, 2nd edn. (Wiley, Hoboken, 2015). ISBN: 978-1-119- 00954-2. https://doi.org/10.1002/9781119009597
33. R.W. Erickson, D. Maksimović, *Fundamentals of Power Electronics*, 2nd ed. (Springer, 2001). https://doi.org/10.1007/b100747
34. D. Czarkowski, M.K. Kazimierczuk, Static- and dynamic-circuit models of PWM buck-derived DC-DC convertors. IEE Proc. G - Circuits Devices Syst. **139**(6), 669–679 (1992). https://doi.org/10.1049/ip-g-2.1992.0101
35. R.W. De Doncker, D.M. Divan, M.H. Kheraluwala, A three- phase soft-switched high-power-density DC/DC converter for high power applications. IEEE Trans. Ind. Appl. **27**(1), 63–73 (1991). https://doi.org/10.1109/28.67533
36. H. Yuan, K. Li, S. Li, S.-C. Tan, S.Y.R. Hui, On the control of DC-DC converters in the SS-compensated wireless power transfer system, in *Proceedings of the 2024 IEEE Wireless Power Technology Conference and Expo (WPTCE)*, Kyoto (2024), pp. 788–792. https://doi.org/10.1109/wptce59894.2024.10557334
37. Y. Yang, J. Zeng, S.Y.R. Hui, A fast primary-side current and voltage control for direct wireless battery chargers. IEEE J. Emerg. Sel. Top. Power Electron. **12**(1), 355–366 (2024). https://doi.org/10.1109/jestpe.2023.3316134
38. Y. Mao, Y. Yang, K. Wang, A primary-side fixed-frequency CC and CV output control for single-stage wireless battery chargers with series compensation receivers. IEEE Trans. Transp. Electrif. **10**(2), 3407–3415 (2024). https://doi.org/10.1109/tte.2023.3297573

Chapter 5
Applications and Future Trends

This chapter examines representative applications and future trends in wireless power transfer (WPT), focusing on high-voltage transmission-line monitoring and electric-vehicle charging. These case studies show how converter-topology and compensation-network selection depend on application-specific constraints, operating power levels, and practical requirements such as cost, interoperability, and standardization. The chapter further discusses emerging directions in WPT, including dynamic charging, consumer ecosystems, biomedical and wearable devices, specialty vehicles, and the enabling roles of wide-bandgap devices, sensing, and embedded control. It also highlights tightening requirements related to safety, EMC/EMI, and certification. The main conclusion is that topology selection is now a system-level task in which converters, resonant networks, couplers, control methods, and compliance constraints must be considered jointly.

5.1 Infrastructure and Transportation Applications

This section focuses on two representative application domains at opposite ends of the deployment spectrum: a niche infrastructure application in high-voltage transmission-line monitoring and a mass-market transportation application in electric-vehicle charging. These examples show how converter and compensation-topology choices depend on application-specific constraints, power levels, and practical considerations such as cost and standardization. Other wireless power transfer applications have also been widely reported in the literature, including biomedical implants, personal wearables, and specialty vehicles such as electric bicycles, unmanned aerial vehicles (UAVs), and autonomous underwater vehicles (AUVs).

D. Czarkowski et al., *Converter Topologies for Resonant Wireless Power Transfer*, Power Systems, https://doi.org/10.1007/978-3-032-31448-2_5

5.1.1 WPT for High-Voltage Power Lines

As a novel application of wireless power transfer (WPT), this subsection addresses the problem of reliable power supply for online monitoring terminal devices (OMTDs) installed on high-voltage (HV) power transmission lines. These OMTDs are important Internet-of-Things components of a smart grid (Fig. 5.1), where they support information sharing and intelligent control, thereby improving system reliability [1–3].

Reliable and maintenance-free power delivery remains challenging for such outdoor terminal devices. Conventional solar sources are strongly weather dependent and require periodic maintenance. Rogowski-coil-based schemes have been proposed to harvest power from overhead current-carrying conductors [4], but those coils remain at the same potential as the monitored line and therefore can only supply devices mounted close to the conductor. For the many grounded terminal devices installed on the tower, the HV electrical power cannot be used directly because the required insulation must be preserved. Consequently, a reliable and maintenance-free supply solution is still needed.

WPT based on resonant coupling provides an attractive alternative. A practical HV WPT system comprises five main parts: a current-transformer (CT) device,

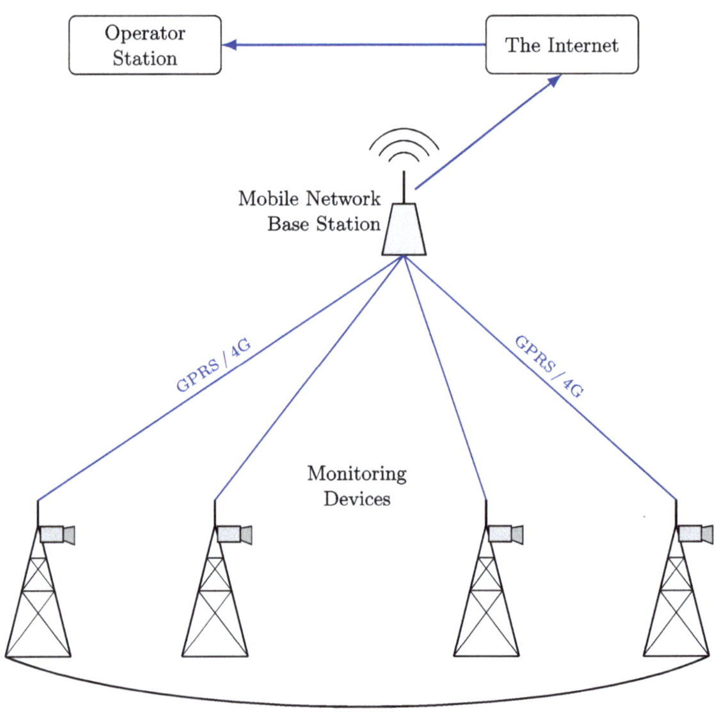

Fig. 5.1 Smart grid with the Internet of Things

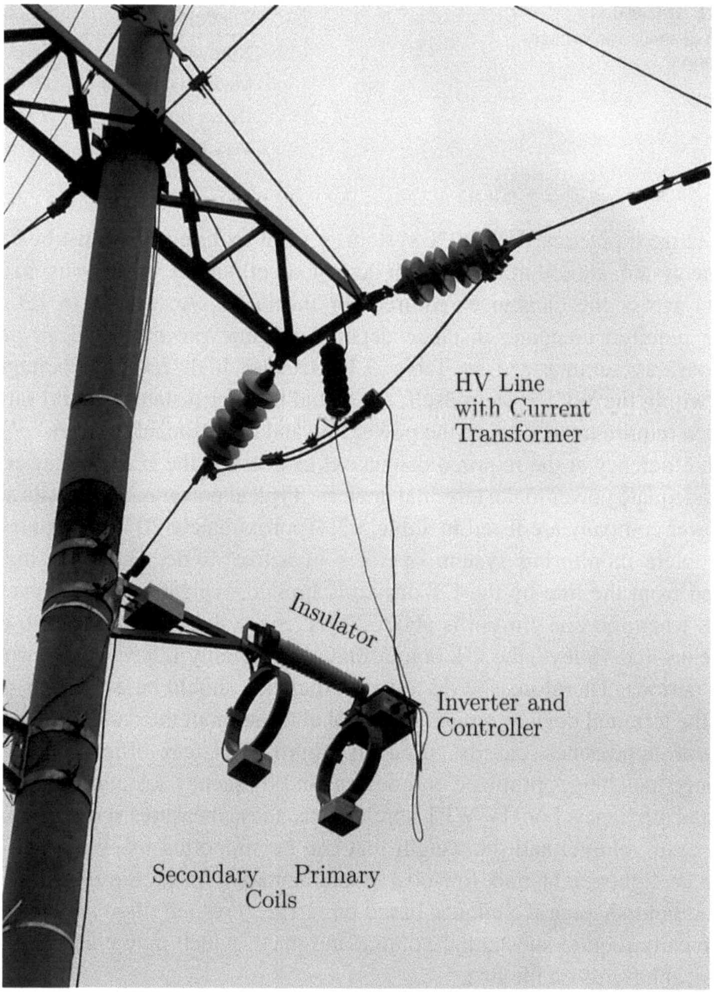

Fig. 5.2 Overview of HV WPT applications

a high-frequency inverter, coupled LC resonators, a rectifier, and the load. In a representative implementation demonstrated on the Guangxi power grid in China (Fig. 5.2), the CT device, inverter, and transmitting resonator are mounted close to the line and remain at line potential, whereas the receiving-side components and the load are mounted on the tower and grounded for improved electromagnetic compatibility. In this way, electrical power is first extracted from the transmission line by the CT device and then transferred wirelessly from the HV conductor to the tower, so that the load can be powered without compromising insulation between the line and the tower. Since WPT can operate under a wide range of weather conditions, it can serve as a highly reliable auxiliary source.

Table 5.1 IEC 60137 creepage distance for voltages above 1000 V

Condition	Creepage
Normal	$16\,\text{mm}\,\text{kV}^{-1}$
Medium pollution	$20\,\text{mm}\,\text{kV}^{-1}$
Heavy pollution	$25\,\text{mm}\,\text{kV}^{-1}$
Very heavy pollution	$31\,\text{mm}\,\text{kV}^{-1}$

To realize a practical HV WPT system, several requirements must be satisfied. First, the system should maintain high dc-to-load efficiency while transferring tens of watts across the clearance required for insulation. According to IEC 60137 [5], the required creepage distance depends on line potential and air-pollution conditions, as summarized in Table 5.1. Because high resonant voltages also appear within the WPT system itself, the actual transfer distance should satisfy the insulation requirements of both the power line and the resonant network.

High efficiency at the required distance is essential if the HV WPT system is to supply adequate power to the terminal devices. Typical power requirements supplied by a power company are listed in Table 5.2. Approximately 20 W is required when the complete monitoring system operates in active mode. However, the power extracted from the line by the CT device is limited, typically to only several tens of watts when the line current is about 200 A. Since the line current also exhibits daily peaks and valleys, the CT output may drop to only a few watts during low-current periods. Therefore, the dc-to-load efficiency should be as high as possible so that the terminal devices remain operational throughout the daily current cycle.

Several approaches can be used to improve system efficiency, including impedance matching, optimized coil design, and frequency selection for maximum dc-to-load efficiency. For HV WPT applications, these measures should be balanced against cost, volume, and the weight that can be supported by the power line. In practice, a lightweight and low-cost implementation is preferred. Accordingly, magnetic field-shaping techniques based on ferrite cores are often avoided because they typically require substantial volume and mass, which may create undesirable mechanical hazards on the line.

Second, the electromagnetic-interference (EMI) issues of the HV WPT system must be addressed. On the transmitting side, the control circuitry and high-frequency inverter are mounted close to the power line and therefore must withstand the strong electromagnetic field produced by the line. This is commonly handled by shielding. If the shielding enclosure and the analog ground of the transmitting

Table 5.2 Typical power requirements of online monitoring systems (courtesy of Guangdong Power Grid Corporation)

Subsystem	Sleep mode	Active mode
Video	2 W	12 W
Snow monitoring	1 W	3.5 W
Micro-meteorological monitor	1 W	3 W
Micro-processing unit	10 mW	50 mW
Total	4.01 W	18.55 W

circuits are connected to the line potential, the electronics within the enclosure share the line potential and can achieve satisfactory electromagnetic compatibility. On the receiving side, the monitoring devices must be protected both from radiated interference produced by the WPT system and from conducted interference along the circuit interconnections. Since the receiving-side circuits can be grounded and the monitoring devices can be placed away from the receiving coil, EMI is usually more manageable there. In addition, the magnetic field produced by the line current has little effect on the WPT link because the line-frequency field is far removed from the several-hundred-kilohertz to megahertz operating range of the HV WPT system.

Third, because the HV WPT system is an outdoor installation, long-term stability and reliability are of major importance. The inverter should operate efficiently and be designed for adequate thermal management; otherwise, direct solar radiation may increase the temperature inside the shielding enclosure to more than 60 °C in summer and cause failure. The natural frequencies of the transmitting and receiving resonators may also drift because of humidity, temperature, and stray coupling to nearby objects, and such drift can severely reduce both output power and efficiency. Conventional tuning with inductor–capacitor arrays is possible, but the transmitting circuits are typically shielded and communication between the transmitting and receiving sides is difficult to establish. For this reason, the tuning network is preferably placed on the transmitting side, and the tuning algorithm should rely only on transmitting-side information.

Before a prototype can be deployed on HV power lines, several electromagnetic-compatibility and insulation tests must be passed, including lightning tests, dielectric withstand-voltage tests, and power-frequency magnetic-field immunity tests. The design should also account for rain, wind loading, and mechanical vibration of the transmission line.

In summary, WPT offers a promising solution to the problem of reliable power supply for online monitoring devices on HV transmission lines. Practical realization, however, still requires careful system-level design. To date, only limited field-oriented implementations have been reported for transmission-line monitoring in high-voltage environments, including magnetically resonant wireless power systems developed for monitoring terminals and studies of the electromagnetic behavior of WPT systems deployed on transmission lines [6, 7]. Nevertheless, the application potential is considerable.

With these constraints in mind, a practical system can be outlined as follows. The topology of a practical HV WPT system is shown in Fig. 5.3. The system includes the CT device with its management circuitry, the high-frequency inverter, the coupled resonators, the rectifier, and the load.

The electrical power harvested by the CT device is 50 Hz AC, represented by V_{CT} in Fig. 5.3. In principle, V_{CT} can be converted directly to high-frequency AC by the single-phase AC/AC converter shown in Fig. 5.4. In practice, however, the AC/AC approach leads to large current ripple in the resonant tank during energy oscillation, which complicates output-power regulation. A more practical approach is to rectify V_{CT} at power frequency, smooth it with a capacitor C_{CT}, and then

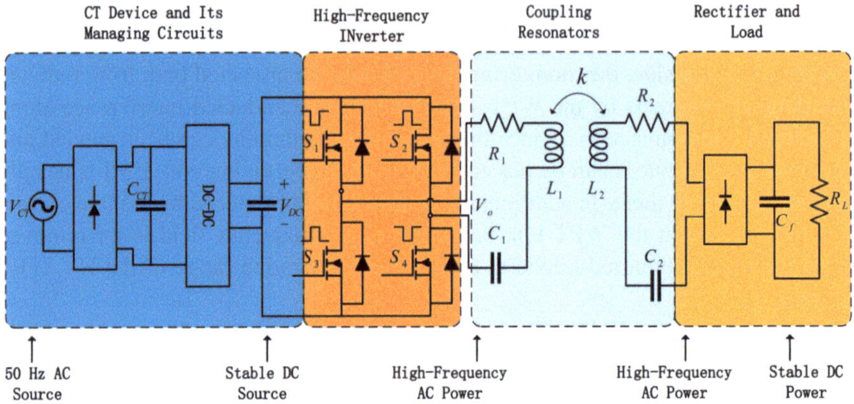

Fig. 5.3 Overall circuit diagram for HV WPT systems

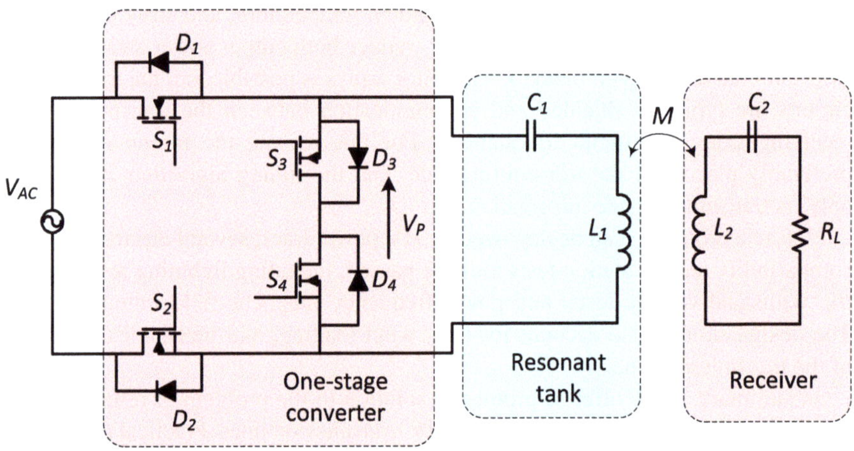

Fig. 5.4 Single-phase AC/AC converter

generate a regulated DC source by means of a DC/DC module. Since current-transformer power-harvesting modules are already fairly mature [8], the following discussion emphasizes the wireless power-transfer portion of the system.

The high-frequency inverter is a key element of the HV WPT application, and its design should account for heat dissipation, power level, and operating-frequency range. For an efficient WPT system spanning a large transfer gap, for example 1.5 m for 110 kV power lines, the operating frequency typically lies between several hundred kilohertz and a few megahertz. If the dc-to-load efficiency at a distance of 1.5 m is only 20 %, roughly 100 W must be transmitted in order to deliver 20 W to the load. In addition, because active cooling is generally not available on the transmission line, the loss in each transistor must remain within the thermal limit of the installation.

Table 5.3 Characteristics of different high-frequency power supplies for HV WPT applications

Topology	Switch number	Power level	Frequency	Typical application
Class-E	Single	<10 W	>1 MHz	Artificial organs, low-power systems
Class-D half-bridge	Double	1–3000 W	100–1000 kHz	Medium-power systems
Class-D full-bridge	Multiple	All levels	20–1000 kHz	Electric vehicles, chargers, and related systems

Common driver topologies include Class-D and Class-E inverters. Their basic characteristics are compared in Table 5.3. Relative to the Class-E inverter, the Class-D inverter, especially in the full-bridge configuration, offers higher power capability and simpler heat dissipation. For this reason, the full-bridge Class-D inverter is usually chosen for HV WPT applications. If additional space and mass can be tolerated, two full-bridge Class-D inverters may also be connected in parallel to obtain higher and more flexible output power as well as improved thermal distribution [9]. This arrangement retains a degree of fault tolerance if one bridge experiences an open-circuit fault, although a short-circuit fault still requires the entire system to be shut down. The corresponding parallel-inverter concept is shown in Fig. 5.5.

For practical operation, the inverter is often driven under inductive-load conditions so that turn-on switching loss is minimized. Since turn-off loss still remains, MOSFETs with very short fall times are preferred. For example, the practical system under discussion uses CSD19534KCS MOSFETs to implement the Class-D full-bridge inverter, and their 1 ns fall times help reduce turn-off loss.

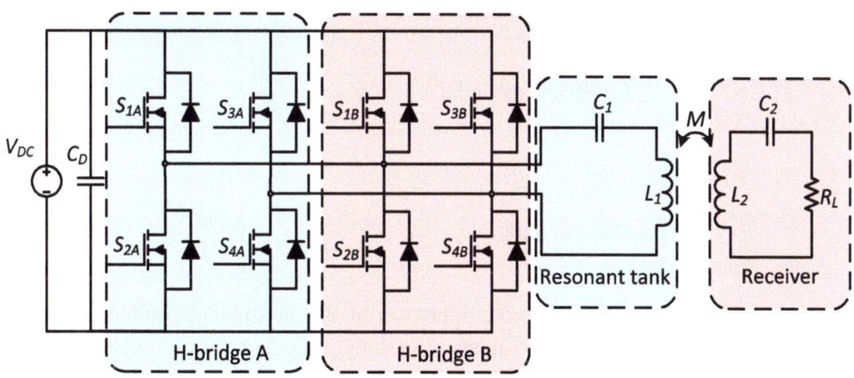

Fig. 5.5 Parallel full-bridge Class-D resonant inverter

Apart from the driver circuit, the compensation topology of the resonators is also essential to good performance. Four basic topologies exist, depending on the placement of the compensation capacitors at the transmitting and receiving coils: series–series (SS), series–parallel (SP), parallel–series (PS), and parallel–parallel (PP) [10, 11]. On the secondary side, the series topology leads to a reflected impedance

$$Z_r = \frac{\omega^2 M^2}{Z_s} = \frac{\omega^2 M^2}{j\omega L_2 + \dfrac{1}{j\omega C_2} + R_2 + R_L}, \tag{5.1}$$

whereas the parallel topology yields

$$Z_r = \frac{\omega^2 M^2}{Z_s} = \frac{\omega^2 M^2}{j\omega L_2 + R_2 + \dfrac{1}{j\omega C_2 + 1/R_L}}. \tag{5.2}$$

Here, ω is the angular operating frequency and M is the mutual inductance related to the magnetic coupling factor k by

$$M = k\sqrt{L_1 L_2}. \tag{5.3}$$

Usually, the operating frequency is chosen equal to the resonant frequency of the secondary side,

$$\omega_0 = \frac{1}{\sqrt{L_2 C_2}}.$$

Then (5.1) and (5.2) can be rewritten as

$$\text{for series-secondary:} \qquad Z_r = \frac{\omega_0^2 M^2}{R_L}, \tag{5.4}$$

$$\text{for parallel-secondary:} \qquad Z_r = \frac{R_L M^2}{L_2^2} - j\,\frac{\omega_0 M^2}{L_2}. \tag{5.5}$$

As indicated by (5.4) and (5.5), operation at the nominal secondary resonant frequency makes the series-secondary topology reflect a purely resistive load, whereas the parallel-secondary topology reflects an additional imaginary term related to the mutual inductance. Consequently, mutual-inductance variation shifts the nominal resonant frequency seen at the primary side in the parallel case, and the

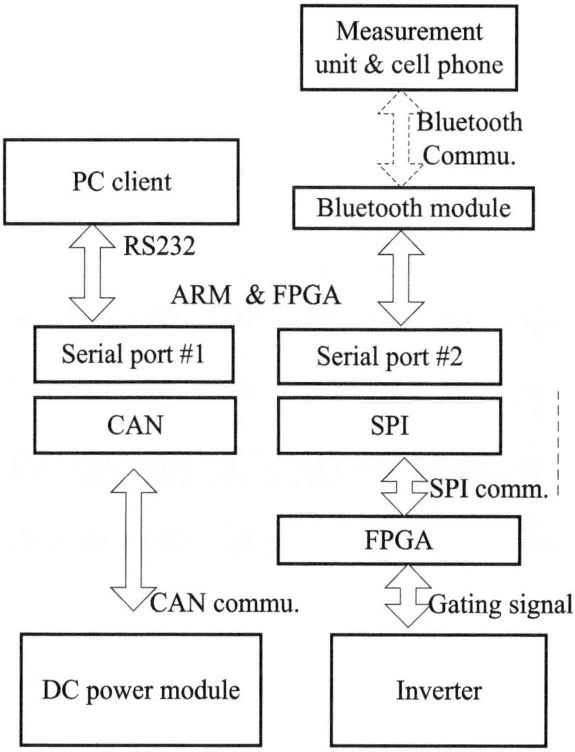

Fig. 5.10 Schematic of the control unit

among the transmitting-side controller, the receiving-side measurement unit, and a mobile device. Through a CAN bus connected to the DC module, the controller regulates the module output voltage according to the battery charging voltage and current measured on the secondary side.

The control unit contains two microcontrollers, an ARM device and an FPGA, along with supporting peripherals [27]. The ARM STM32F407 provides three serial communication ports, including interfaces for an RS232 link to a PC client and for a Bluetooth module. CAN and SPI interfaces are used to communicate with the DC power module and FPGA, respectively (Fig. 5.10).

In practice, charging an EV battery pack proceeds through three stages: constant-current charging, constant-voltage charging, and trickle charging. A typical charging trajectory is shown conceptually in Fig. 5.11, and 17 measured operating points are listed in Table 5.5. The equivalent load is

$$R_{\text{Load}} = \frac{V_{\text{Load}}}{I_{\text{Load}}}, \tag{5.6}$$

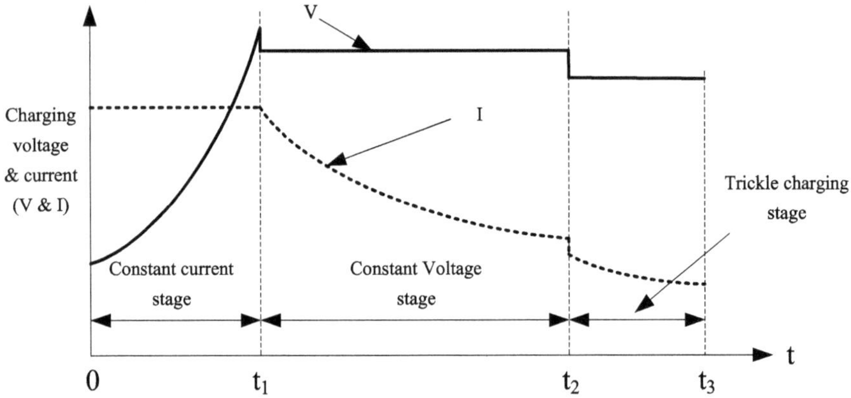

Fig. 5.11 Typical charging process with three stages (constant current, constant voltage, and trickle charging)

Table 5.5 Measured charging parameters from a cable-charging process

Point	V_{Load} (V)	I_{Load} (A)	R_{Load} (Ω)	P_{Load} (W)
1	140.3	20.04	7.00	2812
2	145.5	19.99	7.28	2908
3	148.3	19.89	7.46	2950
4	150.4	19.85	7.58	2986
5	152.8	19.80	7.72	3026
6	156.0	19.31	8.08	3012
7	157.4	19.10	8.24	3006
8	159.6	18.96	8.42	3026
9	160.6	18.84	8.52	3026
10	162.8	18.55	8.78	3020
11	165.8	18.10	9.16	3000
12	171.7	15.24	11.27	2616
13	172.6	10.65	16.21	1838
14	172.7	8.19	21.09	1414
15	173.0	6.98	24.79	1208
16	173.2	6.12	28.30	1060
17	173.3	5.70	30.40	988

In the following discussion, the Class-D half-bridge inverter and the SS compensation network are assumed. Using the relations developed earlier for the Class-D half-bridge, the required input DC voltage of the DC power module can be written as

$$V_{DC} = \frac{\pi P_1}{I_p \cos\psi}, \tag{5.7}$$

where P_1 is the total power delivered by the inverter per

$$P_1 = \frac{1}{2}\left(I_p^2 r_p + I_s^2 r_s\right) + P_{\text{Load}}, \tag{5.8}$$

I_p is the primary-side current amplitude, and ψ is the phase angle by which the current lags the inverter voltage. The current amplitude is

$$I_p = \left|\frac{Z_i I_s}{\omega M}\right| = \left|\frac{\left(j\omega L_s + 1/(j\omega C_s) + 8R_{\text{Load}}/\pi^2 + r_s\right)\pi I_{\text{Load}}/2}{\omega M}\right|, \tag{5.9}$$

where L_s and C_s are the secondary-side resonant inductance and capacitance. The phase angle is

$$\psi = \arctan\left(\frac{\text{imag}(Z)}{\text{real}(Z)}\right), \tag{5.10}$$

and the inverter input impedance is

$$Z = Z_1 + Z_{\text{ref}} = j\omega L_p + \frac{1}{j\omega C_p} + r_p + \frac{(\omega M)^2}{r_s + j\omega L_s + 1/(j\omega C_s) + 8R_{\text{Load}}/\pi^2}. \tag{5.11}$$

Here, L_p and C_p are the primary-side resonant inductance and capacitance, and r_p and r_s are the primary- and secondary-side equivalent series resistances (ESRs), respectively. In the numerical example, they are assumed equal for simplicity.

Under the parameter set

$$\begin{aligned} &L_p = L_s = 37\,\mu\text{H}, \qquad M = 7.33\,\mu\text{H}, \qquad C_p = C_s = 105\,\text{nF}, \\ &\omega = 5.6 \times 10^5\,\text{rad s}^{-1}, \quad r_p = r_s = 0.075\,\Omega \end{aligned} \tag{5.12}$$

the required input DC voltage is shown in Fig. 5.12. This plot indicates that the REG75030 module, with output voltage adjustable from 50 to 750 V, satisfies the required voltage range. A corresponding laboratory setup is shown in Fig. 5.13.

5.2 Future Trends

Future trends in wireless power transfer (WPT) are shaped by application pull (EVs, consumer ecosystems, medical implants, robotics), technology push (wide-bandgap devices, sensing, embedded control), and increasingly tight constraints related to safety, EMC/EMI, and interoperability. From the standpoint of this monograph, the central observation is that topology selection is no longer a stand-alone task: *converter choices, resonant networks, couplers, sensing, and certification*

Fig. 5.12 Required DC-module output voltage versus charging voltage

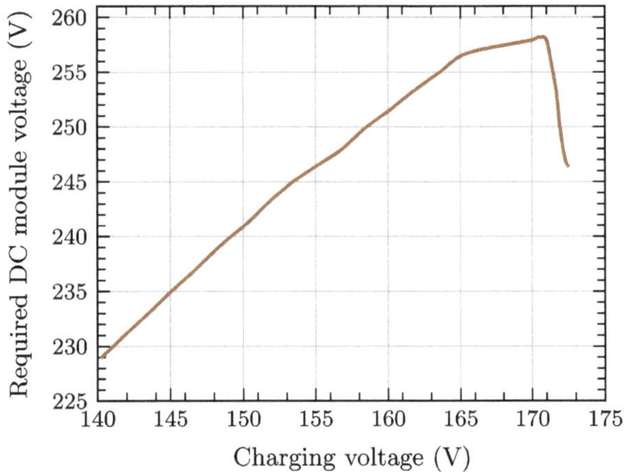

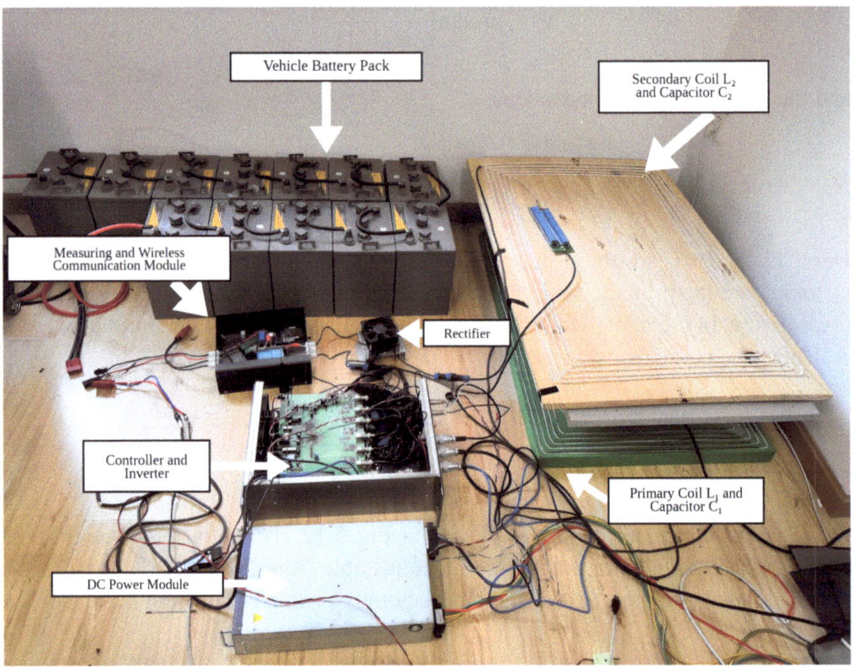

Fig. 5.13 Laboratory prototype of the EV wireless charging setup

constraints increasingly co-determine one another. At the same time, practical system development is moving away from long conversion chains with multiple isolated optimization targets and toward reduced-stage architectures in which the front-end, resonant link, and output conditioning are more tightly integrated. This trend is especially visible in recent work on matrix converters, bidirectional V2Gcapable systems, multi-level power stages, and direct AC/AC wireless charging structures.

Recent standards activity also includes SAE J2954/3 for dynamic wireless power transfer, while the WPC ecosystem continues to evolve through Qi2 and Qi2 25W alongside Ki.

5.2.1 *Trend Drivers and Their Implications for Resonant Converters*

Three recurring macro-drivers emerge across standards and system-level surveys:

- **Interoperability at scale:** consumer (Qi/Qi2 and Ki) and automotive (SAE J2954) ecosystems require predictable behavior across manufacturers, which pushes convergence in operating bands, negotiation/handshakes, and safety tests [24, 28, 29].
- **Operation under uncertainty:** misalignment, variable coupling, and multi-receiver scenarios demand high efficiency over a wide impedance envelope rather than at a single operating point [30].
- **Safety- and EMC-first design:** exposure limits and foreign-object/living-object detection (FOD/LOD) influence allowable waveforms, modulation schemes, and sensing integration [31–33].

The net effect is a shift from "pick a topology and optimize it" to *joint design* in which topology, modulation, and feedback sensing are selected together to satisfy efficiency *and* compliance targets.

5.2.2 *From Point Design to Impedance-Envelope Design*

In practical WPT systems, the reflected load seen by the inverter varies with coupling, alignment, and receiver operating state. As a result, converter design is increasingly judged by its *impedance-envelope performance*: the ability to maintain soft switching and controllable power flow while the effective load trajectory sweeps a large region of the complex plane. This favors resonant families that offer: (i) inherent or controllable soft switching (ZVS/ZCS), (ii) modulation freedom (frequency/phase/duty control), and (iii) tractable parameter identification for robust closed-loop operation. These requirements bear directly on the topologies emphasized elsewhere in this monograph (Chaps. 2 and 3), because the "best" resonant network and the "best" inverter often depend on the same impedance envelope and sensing assumptions. The same perspective also helps explain growing interest in reduced-stage conversion. In conventional EV-oriented WPT, energy often passes through AC/DC, DC/AC, wireless AC/AC, and then AC/DC or DC/DC stages before reaching the battery. Each additional stage can improve local controllability, but it also adds switching loss, volume, cost, and control overhead. Future topologies are therefore judged increasingly not only by peak resonant-link efficiency, but also by how effectively they collapse or integrate stages across the full power path. This same systems perspective motivates the topology directions discussed next, which aim not only to preserve soft switching over a wide operating envelope but also to simplify the end-to-end conversion path.

5.2.3 Higher Frequency Operation and Wide-Bandgap Devices

Wide-bandgap (WBG) devices (GaN, SiC) continue to reduce switching loss and enable higher-frequency operation with smaller passives, but higher frequency also tightens EMI constraints and increases sensitivity to parasitics. In consumer ecosystems, resonant standards and ecosystems have explored ISM-band operation such as 6.78 MHz [34], and multi-MHz resonant inverters (including Class E variants) become appealing because device capacitances can be absorbed into intended resonant operation while preserving soft switching. At the same time, the move toward higher frequency strengthens the case for modeling workflows that are explicitly parasitic-aware and that connect circuit-level choices to measured EMI behavior [30].

5.2.4 EV WPT: Dynamic Charging, Heavy-Duty Scaling, and Infrastructure Integration

For light-duty EV charging, SAE J2954 has pushed the ecosystem toward interoperable operation and alignment methodologies [24]. For higher powers and heavy-duty charging, surveys emphasize the role of the nominal 85 kHz band, magnetic shielding, and thermal constraints [35–37]. A major future direction is **dynamic wireless power transfer** (DWPT), in which roadway infrastructure energizes vehicle receivers in motion. From a converter-topology viewpoint, DWPT strengthens the importance of:

- **modular inverter architectures** (segmented pads, distributed inverters, polyphase excitation);
- **fast transient control** as coupling changes rapidly when pads enter/exit the active region;
- **grid-side coordination** for power quality under stochastic and spatially distributed loading.

Recent EV-focused reviews also highlight that deployment is shaped as much by economics and standardization as by physics, which reinforces the need to align topology choices with certification requirements and infrastructure constraints [38].

5.2.5 Bidirectional Wireless Power and Vehicle-to-Grid (V2G)

Bidirectional power flow (vehicle ↔ grid/home) is a growing systems trend as EV penetration increases. Recent literature indicates that bidirectional wireless charging can support grid services provided that power electronics, communication, and protection are co-designed [39]. At the topology level, bidirectionality tends to favor:

- active rectification and reversible DC/DC stages;
- modulation strategies that preserve ZVS/ZCS in both directions (often through phase/frequency control);
- explicit common-mode and EMI strategies due to longer duty cycles and two-way power flow.

The V2G case is important because it tests whether the wireless charger can evolve from a one-way convenience feature into an active grid asset. Prior system studies compatible with the SAE J2954 framework indicate that bidirectionality can be added with relatively modest hardware changes, while maintaining similar efficiency in grid-to-vehicle (G2V) and vehicle-to-grid (V2G) operation at comparable power levels [40]. Representative V2G-oriented WPT topologies retain a full resonant wireless link but surround it with reversible grid-side and battery-side converters, often using series–series compensation together with bridge rectifier/inverter stages and a synchronous buck/boost interface for the battery [41, 42]. A representative example is shown in Fig. 5.14. In such systems, the dominant challenge shifts away from the coupler itself toward multi-loop control coordination: the grid-side converter, resonant link, and battery interface must jointly regulate power flow, maintain soft switching, and satisfy grid-current quality requirements in both directions of operation.

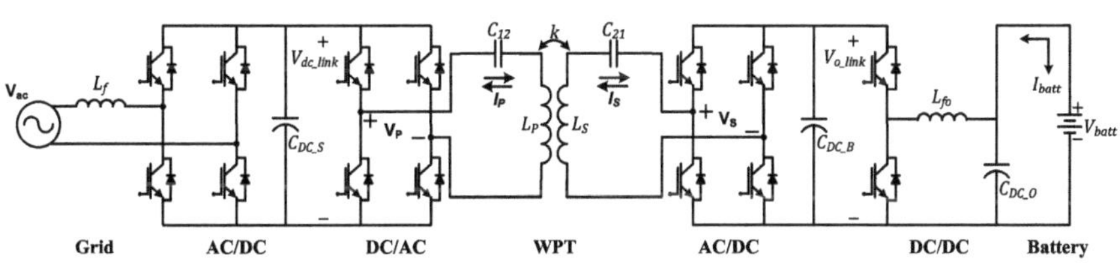

Fig. 5.14 Representative bidirectional WPT topology for vehicle-to-grid operation with a single-phase grid-side converter

5.2.6 Spatial Freedom, Coil Arrays, and Multi-receiver Charging

Consumer and robotics applications increasingly demand *placement freedom* and simultaneous charging of multiple devices. Spatial freedom and multi-device charging are explicitly addressed in [34, 43]. Their converter implications include scalable excitation of coil arrays, receiver identification and power allocation, and robust estimation of coupling and load state for closed-loop control. These requirements tend to promote modular and multi-channel architectures in which each coil (or small coil group) is driven by a controllable power stage rather than a single monolithic inverter.

5.2.7 Safety, Sensing, and "Field Shaping" as First-Class Objectives

As WPT moves into public spaces and higher power levels, safety mechanisms become certifiable requirements rather than optional features. Two directions are particularly relevant to resonant converter design:

1. **Foreign-object/living-object detection (FOD/LOD):** impedance- and perturbation-based methods can be integrated into the power stage, while dedicated sensing coils and thermal approaches provide complementary coverage. Surveys catalog these methods and discuss practical trade-offs in sensitivity and false positives [33]. Recent work also explores data-driven detection (e.g., machine-learning-assisted FOD) as a way to reduce additional sensors in some deployments [44].
2. **Field shaping and shielding:** passive shielding remains dominant in deployed systems, but research continues on engineered structures (metamaterials/metasurfaces) to enhance coupling and/or confine fields. Metamaterial-based enhancement has been demonstrated in prototype studies [45, 46], and review papers summarize design approaches and limitations [47].

Exposure guidelines such as ICNIRP and IEEE C95.1 provide reference frameworks that influence margins, test methodology, and acceptable operating modes [31, 32].

5.2.8 Capacitive and Hybrid WPT: Alternate Couplers and New Design Knobs

Capacitive WPT (CPT) continues to mature as an alternative when magnetic couplers are undesirable, and it is increasingly considered in EV contexts. Reviews highlight the central challenges: plate geometry, stray capacitances, EMC, and

frequency. If these ideas mature, they could become particularly important for cost-sensitive wireless EV charging platforms, where reducing conversion-stage count and bulky passive storage is as valuable as improving peak converter efficiency.

5.2.12 Co-design with Communications, Software, and Verification Workflows

Standardized consumer systems rely on digital communication for negotiation, power control, and safety interlocks. WPC documentation emphasizes defined profiles (e.g., Qi2 and Ki) and certification testing as system-level constraints [28, 29]. For converter and resonant-network designers, this motivates architectures that expose controllable "knobs" (frequency, phase, duty, or impedance tuning) that can be exercised by software to satisfy negotiated power targets while retaining soft switching. In parallel, increasing system complexity encourages *verification workflows* that connect multi-physics constraints (magnetics, thermal, EMF/EMI) to circuit models, enabling rapid iteration and traceability from topology assumptions to compliance evidence [30].

This system-level coupling is also reflected in current standards and recent surveys on interoperability, safety, and field management [24, 28, 32, 33].

5.2.13 Conclusions

The dominant theme is the increasing importance of a systems-level treatment of WPT: converters, resonant networks, couplers, sensing, communications, and standards are inseparable. For resonant WPT in particular, topology decisions are best framed as *conditional* choices tied to the impedance-envelope characteristic, the available sensing for safety and control, and the certification pathway for the targeted application market. These trends are summarized in Table 5.6 and represented graphically in Fig. 5.16. A second major conclusion is that future WPT architectures will increasingly be judged by how efficiently they integrate *the whole conversion chain*, rather than by the resonant transmitter or receiver in isolation. Matrix converters, bidirectional V2G-capable links, multi-level stages, and hybrid AC/AC interfaces all point in the same direction—fewer bulky passive elements, more functionality concentrated into controllable converter stages, and stronger coupling between circuit topology and supervisory control.

Table 5.6 Trend-to-topology mapping: examples of how future requirements propagate into converter choices

Trend	Converter/system implications
Higher frequency, smaller passives	Incentivizes soft-switching resonant inverters (e.g., Class E variants) and parasitic-aware modeling; EMI design becomes a first-order constraint.
Dynamic EV charging (DWPT)	Modular/multi-phase architectures; rapid transient control; pad segmentation; robust operation under fast-changing coupling.
Bidirectional WPT	Active rectification and reversible DC/DC stages; ZVS/ZCS maintenance in both directions; protection and comms co-design [39].
Spatial freedom & multi-device	Coil-array excitation strategies; scalable multi-channel power stages; multi-receiver power allocation and receiver identification.
Safety/EMF constraints	Integrated sensing (FOD/LOD), shielding and field shaping; limits constrain allowable operating modes and modulation strategies [32, 33].
Reduced-stage AC/AC conversion	Matrix converters and hybrid AC/AC stages can eliminate the bulky DC link and improve power density, but they shift design burden to commutation strategy, THD/PF control, and resonant current management [51, 52, 54, 62].
Higher-power WPT scaling	Selective use of parallelized or multi-level converters can reduce current stress and conduction loss, especially in heavy-duty or extreme-fast-charging systems, at the price of increased balancing and supervisory control complexity [55, 56, 59].

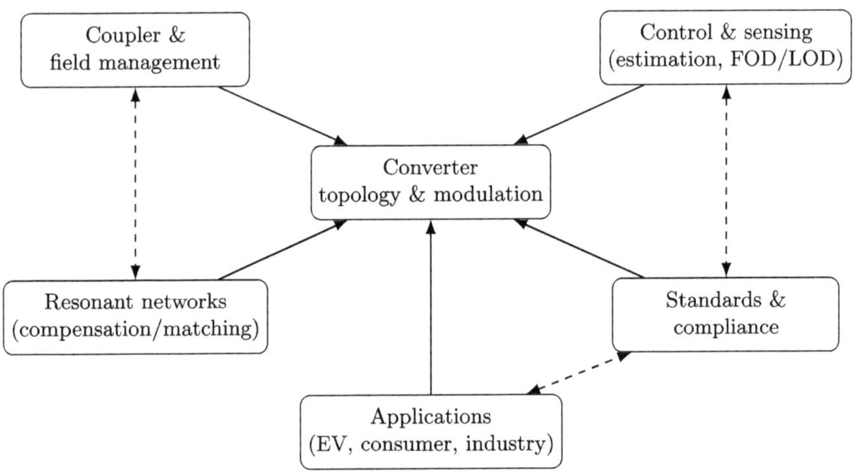

Fig. 5.16 Interdependence of design pressures shaping future WPT systems: converter topology choices increasingly depend on coupler and field constraints, sensing, and certification pathways

References

1. C. Zhang, D. Lin, N. Tang, R.S.Y. Hui, A novel electric insulation string structure with high-voltage insulation and wireless power transfer capabilities. IEEE Trans. Power Electron. **33**(1), 87–96 (2018). https://doi.org/10.1109/tpel.2017.2706221
2. X. Zhang, Y. Zhou, W. Zhou, Z. Dai, M. Li, Space electromagnetic optimization of three-coil WPT for online monitoring equipment of high-voltage transmission lines. IEEE Trans. Electromagn. Compat. **66**(5), 1621–1632 (2024). https://doi.org/10.1109/temc.2024.3417242
3. Y. Liu, Y. Li, X. Zhang, Z. He, Space electromagnetic optimization of three-coil WPT for online monitoring equipment of high-voltage transmission lines. IEEE Trans. Power Electron. **39**(2), 1997–2003 (2024). https://doi.org/10.1109/tpel.2023.3335487
4. L. Du, C. Wang, X. Li, L. Yang, Y. Mi, C. Sun, A novel power supply of online monitoring systems for power transmission lines. IEEE Trans. Ind. Electron. **57**(8), 2889–2895 (2010). https://doi.org/10.1109/tie.2009.2037104
5. International Electrotechnical Commission, Insulated bushings for alternating voltages above 1000 V. IEC 60137 International Standard (2003)
6. L. Gao, X. Gao, H. Zhou, Research and implementation of a WPT system applied in high voltage environment, in *Proceedings of the 2015 International Forum on Energy, Environment Science and Materials* (Atlantis Press, Sep. 2015), pp. 532–538. https://doi.org/10.2991/ifeesm-15.2015.101
7. X. Gao, Q. Deng, W. Hu, H. Zhou, H. Wang, Electromagnetic properties of a WPT system applied on transmission lines, in *2016 Progress in Electromagnetics Research Symposium (PIERS)*, Shanghai (2016), pp. 2787–2790. https://doi.org/10.1109/piers.2016.7735863
8. W. Wang, X. Huang, L. Tan, J. Zhao, C. Yan, Hybrid wireless charging system for monitoring overhead 110 kV high-voltage power line equipment based on magneto-electric conversion. IET Gener. Transm. Distrib. **10**(5), 1199–1208 (2016). https://doi.org/10.1049/iet-gtd.2015.0787
9. A. Schönknecht, R.W. de Doncker, Novel topology for parallel connection of soft-switching high-power high-frequency inverters. IEEE Trans. Ind. Appl. **39**(2), 550–555 (2003). https://doi.org/10.1109/tia.2003.809453
10. A. Khaligh, S. Dusmez, Comprehensive topological analysis of conductive and inductive charging solutions for plug-in electric vehicles. IEEE Trans. Veh. Technol. **61**(8), 3475–3489 (2012). https://doi.org/10.1109/tvt.2012.2213104
11. C.-S. Wang, O.H. Stielau, G.A. Covic, Design considerations for a contactless electric vehicle battery charger. IEEE Trans. Ind. Electron. **52**(5), 1308–1314 (2005). https://doi.org/10.1109/tie.2005.855672
12. A.K.S. Bhat, Analysis and design of LCL-type series resonant converter. IEEE Trans. Ind. Electron. **41**(1), 118–124 (1994). https://doi.org/10.1109/41.281617
13. C. Cai, J. Wang, R. Liu, Z. Fang, P. Zhang, M. Long, M. Hu, Z. Lin, Resonant wireless charging system design for 110-kV high-voltage transmission-line monitoring equipment. IEEE Trans. Ind. Electron. **66**(5), 4118–4129 (2019). https://doi.org/10.1109/tie.2018.2808904
14. J. Ren, P. Hu, D. Yang, D. Liu, Tuning of mid-range wireless power transfer system based on delay-iteration method. IET Power Electron. **9**(8), 1563–1570 (2016). https://doi.org/10.1049/iet-pel.2015.0291
15. D. Patil, M.K. McDonough, J.M. Miller, B. Fahimi, P.T. Balsara, Wireless power transfer for vehicular applications: Overview and challenges. IEEE Trans. Transp. Electrif. **4**(1), 3–37 (2018). https://doi.org/10.1109/tte.2017.2780627
16. A. Ahmad, M.S. Alam, R.C. Chabaan, A comprehensive review of wireless charging technologies for electric vehicles. IEEE Trans. Transp. Electrif. **4**(1), 38–63 (2018). https://doi.org/10.1109/tte.2017.2771619
17. S. Kim, G.A. Covic, J.T. Boys, Tripolar pad for inductive power transfer systems for EV charging. IEEE Trans. Power Electron. **32**(7), 5045–5057 (2017). https://doi.org/10.1109/tpel.2017.2740944

18. T. Kan, T.-D. Nguyen, J.C. White, R.K. Malhan, C.C. Mi, A new integration method for an electric vehicle wireless charging system using LCC compensation topology: analysis and design. IEEE Trans. Power Electron. **32**(2), 1638–1650 (2017). https://doi.org/10.1109/tpel.2016.2552060
19. M. Bojarski, E. Asa, K. Colak, D. Czarkowski, Analysis and control of multiphase inductively coupled resonant converter for wireless electric vehicle charger applications. IEEE Trans. Transp. Electrif. **1**(1), 312–320 (2016). https://doi.org/10.1109/tte.2016.2566921
20. Q. Deng, J. Liu, D. Czarkowski, M. Bojarski, J. Chen, W. Hu, H. Zhou, Edge position detection of on-line charged vehicles with segmental wireless power supply. IEEE Trans. Veh. Technol. **66**(5), 3610–3621 (2017). https://doi.org/10.1109/tvt.2016.2598183
21. A. Kamineni, M.J. Neath, A. Zaheer, G.A. Covic, J.T. Boys, Interoperable EV detection for dynamic wireless charging with existing hardware and free resonance. IEEE Trans. Transp. Electrif. **3**(2), 370–379 (2017). https://doi.org/10.1109/tte.2016.2631607
22. J.M. Miller, O.C. Onar, C.P. White, S.L. Campbell, C. Coomer, L. Seiber, R. Sepe, A. Steyerl, Demonstrating dynamic wireless charging of an electric vehicle: the benefit of electrochemical capacitor smoothing. IEEE Power Electron. Mag. **1**(1), 12–24 (2014). https://doi.org/10.1109/mpel.2014.2300978
23. Z. Li, Q. Deng, S. Li, H. Deng, Q. Huang, W. Hu, LPV modeling and robust control of dynamic wireless charging for electric vehicles. IEEE Trans. Ind. Electron. **72**(10), 10629–10639 (2025). https://doi.org/10.1109/tie.2025.3548993
24. SAE International, *Wireless Power Transfer for Light-duty Plug-in Electric Vehicles and Alignment Methodology* (SAE International, Warrendale, 2024). https://doi.org/10.4271/j2954_202408
25. SAE International, *Wireless Power Transfer for Heavy-duty Electric Vehicles* (SAE International, Dec. 2022). https://doi.org/10.4271/j2954/2_202212
26. J. Sallán, J. Villa, A. Llombart, J.F. Sanz, Optimal design of ICPT systems applied to electric vehicle battery charge. IEEE Trans. Ind. Electron. **56**(6), 2140–2149 (2009). https://doi.org/10.1109/tie.2009.2015359
27. Q. Deng, J. Liu, D. Czarkowski, M.K. Kazimierczuk, M. Bojarski, H. Zhou, Frequency-dependent resistance of Litz-wire square solenoid coils and quality factor optimization for wireless power transfer. IEEE Trans. Ind. Electron. **63**(5), 2825–2837 (2016). https://doi.org/10.1109/tie.2016.2518126
28. Wireless Power Consortium, Qi wireless charging. Accessed 20 Mar 2026 [Online]. Available: https://www.wirelesspowerconsortium.com/standards/qi-wireless-charging/
29. Wireless Power Consortium, Ki cordless kitchen. Accessed 20 Mar 2026 [Online]. Available: https://www.wirelesspowerconsortium.com/standards/ki-cordless-kitchen/
30. J.V. Mulders, D. Delabie, C. Lecluyse, C. Buyle, G. Callebaut, L.V. der Perre, L.D. Strycker, Wireless power transfer: Systems, circuits, standards, and use cases. Sensors **22**(15), 5573 (2022). https://doi.org/10.3390/s22155573
31. International Commission on Non-Ionizing Radiation Protection, Guidelines for Limiting Exposure to Time-Varying Electric and Magnetic Fields (1 Hz to 100 kHz). Health Phys. **99**(6), 818–836 (2010). https://doi.org/10.1097/hp.0b013e3181f06c86
32. IEEE, *IEEE Standard for Safety Levels with Respect to Human Exposure to Electric, Magnetic, and Electromagnetic Fields, 0 Hz to 300 GHz* (Institute of Electrical and Electronics Engineers, 2019). https://doi.org/10.1109/ieeestd.2019.8859679
33. J. Lu, G. Zhu, C.C. Mi, Foreign object detection in wireless power transfer systems. IEEE Trans. Ind. Appl. **58**(1), 1340–1354 (2022). https://doi.org/10.1109/tia.2021.3057603
34. IEEE, *IEEE Standard for Resonant Wireless Power Transfer (WPT) Interface for Information Technology, Communication and Consumer Electronic Equipment* (Institute of Electrical and Electronics Engineers, May 2021). https://doi.org/10.1109/ieeestd.2021.9427471
35. A. Foote, O.C. Onar, A review of high-power wireless power transfer, in *2017 IEEE Transportation Electrification Conference and Expo (ITEC)* (2017), pp. 234–240. https://doi.org/10.1109/itec.2017.7993277

36. H. Feng, R. Tavakoli, Z. Pantic, O. Onar, Advances in high-power wireless charging systems: overview and design considerations. IEEE Trans. Transp. Electrif. **6**(3), 886–904 (2020). https://doi.org/10.1109/tte.2020.3012543
37. C. Panchal, S. Stegen, J. Lu, Review of static and dynamic wireless electric vehicle charging system. Eng. Sci. Technol. Int. J. **21**(5), 922–937 (2018). https://doi.org/10.1016/j.jestch.2018.06.015
38. Z. Xue, W. Liu, C. Liu, K.T. Chau, Critical review of wireless charging technologies for electric vehicles. World Electric Veh. J. **16**(2), 65 (2025). https://doi.org/10.3390/wevj16020065
39. V. Venkatesan, M.M. Azadifar, B. Nahid-Mobarakeh, Bidirectional wireless power transfer: Bridging electric vehicles and the grid. Results Eng. **22**, 102134 (2024). https://doi.org/10.1016/j.rineng.2024.102134
40. K. Tachikawa, M. Kesler, O. Atasoy, Feasibility study of bidirectional wireless charging for vehicle-to-grid, in *SAE Technical Paper 2018-01-0669* (2018). https://doi.org/10.4271/2018-01-0669
41. Y. Yang, M. El Baghdadi, Y. Lan, Y. Benomar, J. Van Mierlo, O. Hegazy, Design methodology, modeling, and comparative study of wireless power transfer systems for electric vehicles. Energies **11**(7), 1716 (2018). https://doi.org/10.3390/en11071716
42. A. Triviño, J.M. González-González, J.A. Aguado, Theoretical analysis of the efficiency of a V2G wireless charger for electric vehicles. Trans. Environ. Electrical Eng. **3**(1), 9 (2018). https://doi.org/10.22149/teee.v3i1.118
43. AirFuel Alliance, *AirFuel Resonant Wireless Power Transfer (WPT) System Baseline System Specification (BSS)* (AirFuel Alliance, Beaverton, 2017) [Online]. Available: https://airfuel.org/standards/resonant-wireless-power/
44. D.Z. Graves, A.V. Bilbao, S.B. Bayne, Machine learning based foreign object detection in wireless power transfer systems. e-Prime – Adv. Electrical Eng. Electron. Energy **7**, 100384 (2024). https://doi.org/10.1016/j.prime.2023.100384
45. B. Wang, K.H. Teo, T. Nishino, W. Yerazunis, J. Barnwell, J. Zhang, Experiments on wireless power transfer with metamaterials. Appl. Phys. Lett. **98**(25), 254101 (2011). https://doi.org/10.1063/1.3595999
46. G. Lipworth, J. Ensworth, K. Seetharam, D. Huang, J.-S. Lee, P. Schmalenberg, T. Nomura, M.S. Reynolds, D.R. Smith, Y. Urzhumov, Magnetic metamaterial superlens for increased range wireless power transfer. Sci. Rep. **4**, 3642 (2014). https://doi.org/10.1038/srep03642
47. W. Lee, Y.-K. Yoon, Wireless power transfer systems using meta-materials: a review. IEEE Access **8**, 147930–147947 (2020). https://doi.org/10.1109/access.2020.3015176
48. W. Zhou, M. Li, C. Zhang, Y. Cao, Z. Wang, J. Yuan, Potential and challenges of capacitive power transfer systems for wireless EV charging: an overview. E-Transportation **20**, 100297 (2024). https://doi.org/10.1016/j.etran.2024.100297
49. J. Zhang, S. Yao, L. Pan, Y. Liu, C. Zhu, A review of capacitive power transfer technology for electric vehicle applications. Electronics **12**(16), 3534 (2023). https://doi.org/10.3390/electronics12163534
50. Z. Dai, M. Sun, H. Chen, X. Ding, H. Wang, Y. Zhai, J. Song, S. Ye, M. Zhang, X. Zhang, Hybrid wireless power transfer: a review. Wirel. Power Transf. **12**, e017 (2025). https://doi.org/10.48130/wpt-0025-0014
51. H.L. Li, A.P. Hu, G.A. Covic, A direct AC–AC converter for inductive power-transfer systems. IEEE Trans. Power Electron. **27**(2), 661–668 (2012). https://doi.org/10.1109/tpel.2011.2159397
52. S. Samanta, A.K. Rathore, A new inductive power transfer topology using direct AC–AC converter with active source current waveshaping. IEEE Trans. Power Electron. **33**(7), 5565–5577 (2018). https://doi.org/10.1109/tpel.2017.2750081
53. S. Weerasinghe, U.K. Madawala, D.J. Thrimawithana, A matrix converter based bidirectional contactless grid interface. IEEE Trans. Power Electron. **32**(3), 1755–1766 (2017). https://doi.org/10.1109/tpel.2016.2550228

54. M. Moghaddami, A.I. Sarwat, Single-phase soft-switched AC–AC matrix converter with power controller for bidirectional inductive power transfer systems. IEEE Trans. Ind. Appl. **54**(4), 3760–3770 (2018). https://doi.org/10.1109/tia.2018.2819695
55. E. Agostini, I. Barbi, Three-phase three-level PWM DC–DC converter. IEEE Trans. Power Electron. **26**(7), 1847–1856 (2011). https://doi.org/10.1109/tpel.2010.2103940
56. E. Asa, O.C. Onar, J. Pries, V.P. Galigekere, G.-J. Su, A trade-off analysis of series/parallel three-phase converter topologies for wireless extreme chargers, in *2020 IEEE Transportation Electrification Conference and Expo (ITEC)* (2020), pp. 1093–1101. https://doi.org/10.1109/itec48692.2020.9161592
57. I. Subotić, N. Bodo, E. Levi, Single-phase on-board integrated battery chargers for EVs based on multiphase machines. IEEE Trans. Power Electron. **31**(9), 6511–6523 (2016). https://doi.org/10.1109/tpel.2015.2497673
58. D. Karwatzki, A. Mertens, Generalized control approach for a class of modular multilevel converter topologies. IEEE Trans. Power Electron. **33**(4), 2888–2900 (2018). https://doi.org/10.1109/tpel.2017.2690226
59. A. Fontanillas Mart'inez, S. Busquets-Monge, J. Bordonau, Modulation and capacitor voltage balancing control of multilevel NPC dual active bridge DC–DC converters. IEEE Trans. Ind. Electron. **67**(4), 2499–2510 (2020). https://doi.org/10.1109/tie.2019.2914643
60. A. Stupar, T. McRae, N. Vukadinović, A. Prodić, J.A. Taylor, Multi-objective optimization of multi-level DC–DC converters using geometric programming. IEEE Trans. Power Electron. **34**(12), 11912–11939 (2019). https://doi.org/10.1109/tpel.2019.2896962
61. P.S. Huynh, D. Vincent, L. Patnaik, S.S. Williamson, FPGA-based PWM implementation of matrix converter in inductive wireless power transfer systems, in *2018 IEEE PELS Workshop on Emerging Technologies: Wireless Power Transfer (WoW)* (2018), pp. 1–6. https://doi.org/10.1109/wow.2018.8450935
62. E. Asa, J. Pries, V.P. Galigekere, S. Mukherjee, O.C. Onar, G.-J. Su, B. Ozpineci, A novel AC–AC wireless power transfer system for EV charging applications, in *2020 IEEE Applied Power Electronics Conference and Exposition (APEC)* (2020), pp. 1685–1690. https://doi.org/10.1109/apec39645.2020.9124602
63. O.C. Onar, M. Chinthavali, S.L. Campbell, L.E. Seiber, C.P. White, V.P. Galigekere, Modeling, simulation, and experimental verification of a 20-kW series-series wireless power transfer system for a toyota RAV4 electric vehicle, in *2018 IEEE Transportation Electrification Conference and Expo (ITEC)* (2018), pp. 436–441. https://doi.org/10.1109/itec.2018.8450085

MIX
Papier aus verantwortungsvollen Quellen
Paper from responsible sources
FSC® C105338

If you have any concerns about our products,
you can contact us on
ProductSafety@springernature.com

In case Publisher is established outside the EU,
the EU authorized representative is:
Springer Nature Customer Service Center GmbH
Europaplatz 3, 69115 Heidelberg, Germany

Printed by Libri Plureos GmbH
in Hamburg, Germany